THE WORLD OF SCIENCE

THE WORLD
BENEATH US

THE WORLD OF SCIENCE
THE WORLD BENEATH US

ANITA McCONNELL

Facts On File Publications
New York, New York ● Bicester, England

THE WORLD BENEATH US

Copyright © 1985 by Orbis Publishing Limited,
London

First published in the United States of America in
1985 by Facts on File, Inc., 460 Park Avenue South,
New York, N.Y. 10016

First published in Great Britain in 1985 by Orbis
Publishing Limited, London

**Library of Congress Cataloging in Publication
Data**

Main entry under title:

World of Science

   Includes index.
   Summary: A twenty-five volume encyclopedia of
scientific subjects, designed for eight- to-twelve-year-
olds. One volume is entirely devoted to projects.
   1. Science—Dictionaries, Juvenile.   1. Science—
Dictionaries
Q121.J86   1984   500   84-1654

ISBN: 0-8160-1068-4

Printed in Italy
10 9 8 7 6 5 4 3 2 1

Consultant editors
Eleanor Felder, Former Managing Editor, *New Book
of Knowledge*
James Neujahr, Dean of the School of Education, City
College of New York
Ethan Signer, Professor of Biology, Massachusetts
Institute of Technology
J. Tuzo Wilson, Director General, Ontario Science
Centre

Previous pages
Stalagmites and
stalactites in the
spectacular Lubang
Angin cave on the
island of Sarawak in
south-east Asia.

Editor Penny Clarke
Designer Roger Kohn

Note There are some unusual words in this book. They are explained in the Glossary on pages 60–63. The first time each word is used in the text it is printed in *italics*.

▼ The various colours of these stalactites and stalagmites in the Aven Armand caves in France are due to different types of mineral in the rocks through which water has seeped.

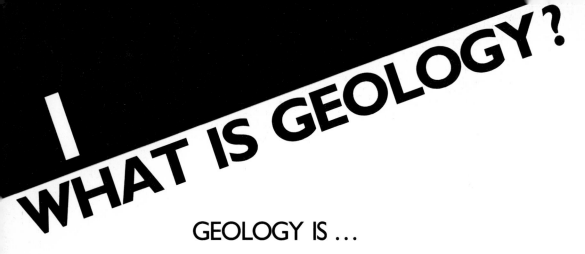

WHAT IS GEOLOGY?

GEOLOGY IS ...

The study of the Earth beneath us is a special science called 'geology', from the ancient Greek word 'geo' for earth. Through geology we try to understand the planet on which we live. Geologists (the scientists who make a special study of geology) investigate the Earth's origin and development, its size, shape and composition. They study the processes that have shaped its surface. They look for evidence in the rocks to teach them about the origins and evolution of life.

Reading the past
Most scientists believe that the Earth has existed for over four thousand million years. How can we hope to learn about its history over such a vast span of time? Geologists maintain that 'The present is the key to the past'. This means that they can look at what is happening on the Earth today, and match this to the record of the rocks. For example, we can observe lava pouring from a volcano and spreading in layers around its slopes. If we then find similar layers of lava buried below the Earth's surface, we can be sure that once upon a time, perhaps millions of years ago, volcanoes were erupting there, in just the same way as they are erupting now. Geologists looking at the sea breaking down the cliffs and grinding the fallen stones into sand, believe that the sea has always worked in this way. Therefore if they find similar beds of pebbles and sands, even though they may be hundreds of miles inland, they know they were formed at the foot of ancient sea cliffs.

The importance of minerals
Ours is a mineral civilization. We depend on mineral products from the Earth beneath us for our fuel, for building materials, for metal ores and for many of the other raw materials for our industries. From their knowledge of its history, geologists can suggest where those valuable minerals may be found. They have also devised methods of exploring the Earth's crust to locate them.

Working with other scientists
Geologists use every branch of science in their investigations of the Earth. They work closely with astronomers, to learn what happened in the earliest periods before the formation of the oldest rocks. They call on physicists to help them date events during the Earth's development, when land, air and oceans were very different to what they are today. With biologists they study the evidence of past life on the Earth, including fossils of the many plants and animals that flourished in the distant past and are now long extinct.

▶ It has taken several million years for the Colorado river in Arizona, USA, to cut through the rock to form the deep gorge that we call the Grand Canyon. The rock layers (strata) that show clearly in the photograph formed when the region was under water. Rivers deposited sediments worn away (eroded) from higher ground which gradually built up to form rock. Now the river is, in turn, wearing away (eroding) these rocks as it cuts an ever deeper canyon. The canyon is 349 km (217 miles) long, 6–20 km (4–13 miles) wide and about 1615 m (5300 ft) deep.

THE FORMATION OF THE EARTH

We cannot understand what is happening today at the Earth's surface unless we know what lies beneath it. How and when the solid Earth, oceans and air came into being can also teach us about the origins of life itself. Astronomers believe that our planet condensed from a cloud of dust and gas whirling and eddying around the Sun. All the chemical *elements* that occur naturally today were present when the Earth was first formed.

Rocks formed as the planet solidified from the dust and gas cloud and the interior grew extremely hot. As a result the rocks flowed very slowly. Giant *convection cells* (see illustration page 10) began to stir up this semi-fluid material. Gradually, heavier chemical elements collected at the core while lighter chemical elements formed a mineral crust at the surface. Water and gases were forced out. The lightest gases were lost into space, but eventually water collected in hollows at the surface and an

atmosphere built up which could support life.

The largest proportion of rock is in the *mantle*, between *crust* and *core* (see illustration). It is far denser than the rocks at the surface. Even today, convection cells in the mantle keep the crust moving in great slabs, or plates (page 12). The boundaries between these plates are marked by lines of volcanoes and zones where earthquakes are common.

Studying the invisible Earth
Of course it is quite impossible to send instruments down into the Earth to get samples of the mantle and core. All we can do is to heat tiny samples of surface rock in special furnaces to the same temperature and pressure as the Earth's mantle. Geologists do this to try and learn about what goes on beneath us, but they cannot reproduce the giant convection

▼ Meteorites from space. Some are rich in iron, as if they came from the middle of a planet that has broken up. Others are stony, as if they came from its mantle. The meteorite (**below right**) fell in Kansas, USA. The nickel-iron rock contains crystals of olivine.

cells. The Earth itself must be their laboratory. They explore the Earth with the vibrations sent out by earthquakes. Each time an earthquake jolts the crust these vibrations (known as *seismic waves*) move outwards in all directions and pass right through the Earth. The waves travel faster in the high-density mantle than they do in the lighter crust. When they pass from one type of rock to another they bend, just as light waves are bent when they pass from air to water. Waves reaching the far side of the Earth are recorded on *seismographs* and the time they took to arrive is measured. By comparing many of these records scientists have worked out how far below the surface the core and mantle boundaries are and how the rocks towards the middle of the Earth are denser.

Although we cannot see our own

▲ In a swirlin of dust and ga are forming ro star. This is h astronomers i the creation o Earth and Mo

▼ Sectioned globe showing the position and relative depths of the Earth's inner and outer cores, mantle and crust.

16–40 km (10–25 miles)

2895 km (1800 miles)

6370 km (3960 miles)

3475 km (2160 miles)

1255 km (780 miles)

inner solid core

outer liquid core

mantle

crust

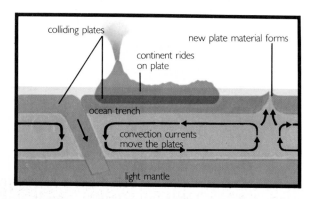

▲ The continental plates and ocean floor are dragged along as deep movements within the Earth, called convection cells, overturn the mantle. Two plates colliding west of South

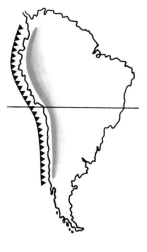

America may cause volcanic eruptions.

planet's core, we can handle part of the core from another planet. The Earth formed at the same time as the rest of the solar system, but some of the other growing planets broke up long ago. Their fragments occasionally reach the Earth as the meteorites which light up the night sky as they come in from space. Geologists now realize that there are two main types of meteorite: stony ones similar to the Earth's crust and mantle, and iron-rich ones that may be similar to the core. For the time being, meteorites that can be analyzed and dated provide the only material which may resemble the middle of the Earth.

HOW GEOLOGISTS EXPLORE THE EARTH

Ordinary methods of mapping just show the Earth's surface features: the mountain ranges, the major plains and valleys. But geologists are just as interested in much more than this. They plot the types of rock visible at the surface. They measure the angle at which the rock layers slope below the ground. In parts of the world such as South Africa, where there are deep mines, they can even venture a few kilometres into the Earth's depths. To probe deeper into the crust they must drill down, either through the sea bed or on dry land, to bring up cores of rock. These cores are then examined under a microscope or by chemical analysis.

In places where we cannot go, we send messengers to travel through the crust. These messengers are small, man-made versions of the vibrations made by earthquakes. Explosive charges are detonated at ground level and send waves down through the rock layers. At each boundary between the different rocks some waves will be reflected back to the surface. Listening posts known as *geophones* record their arrival. By accurately timing all the echoes that return from a single explosion geologists can work out the distance to each rock boundary. From these tests they have also learnt a great deal about the way the Earth has changed over many millions of years.

Geology in the laboratory

Not all geologists work out of doors. A great deal can be learnt from photographs taken from aircraft or by satellites. The bones of the Earth – its geological structure – are clearly visible where there is little soil or vegetation covering the ground. In the laboratory, rocks are sliced and put under the microscope to identify their *minerals*. Chemical analysis shows which elements are present. Rocks can also be dated by measuring their radioactive content. Geologists and biologists work together on the fossils found in some rocks, which show the plants and animals that lived on Earth in ancient times.

▲ The clean, almost polished appearance of the rocks in this cave shaft in the Pennines, England, has been caused by water cutting through the limestone rock over many centuries.

▶ A surveyor at work. Wherever a road is to be built or houses or factories developed surveyors are called in to examine and measure the site.

Measuring gravity

The forces of *gravity* and *geomagnetism* (diagram below left) originating within the Earth, supply information about its mantle and core. Gravity is a fundamental force in our Universe. Everything in the Universe pulls on everything else. The strength of the pull depends on the mass of the objects involved and their distance apart. So the planets revolve round the larger Sun, held in place by the pull of its gravity, and the Moon revolves round the larger Earth. The Earth's gravity holds artificial satellites in orbit, just as it holds the oceans and atmosphere against the Earth's surface.

Within the Earth all matter is attracted towards the middle so that pressure increases from crust to core. The temperature increases with the pressure, melting different rocks at different depths. Because they are not rigid each layer within the Earth moves in slow convective motion. This means that material from the depths wells up to the surface and flows along for a while before sinking again. As a result of these movements, the boundaries between the crust and mantle and the mantle and core are uneven. Because of this unevenness the pull of the Earth's gravity is not equal everywhere. This slight unevenness affects the artificial satellites in their orbits, because where the dense mantle is near to the Earth's surface it pulls the satellites more strongly than where it is under a thicker layer of the Earth's crust.

This slight variation gives *geophysicists* information about the precise shape of the Earth. Scientists have known for several centuries that the Earth is not a true sphere because it is flattened at the poles. Now the shape of satellite orbits show it is very sightly pear-shaped. *Gravimeters*, which are extremely sensitive balances, are set on land or on the sea bed, or even

▲ The Earth's main magnetic field.

▼ A 'solar wind' of electrically charged particles streams past the Earth continually modifying and varying the Earth's magnetic field (magnetosphere). This can be seen by watching a very accurate compass needle, which is moving slightly all the time. The magnetopause is where the magnetic field becomes weak.

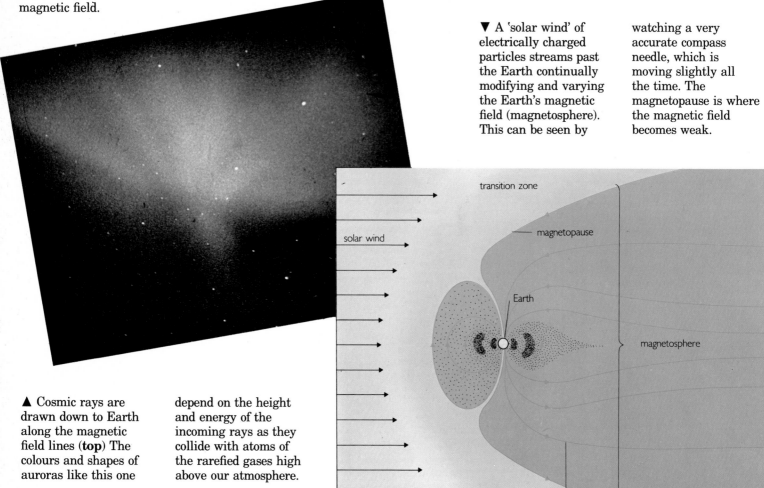

▲ Cosmic rays are drawn down to Earth along the magnetic field lines (**top**) The colours and shapes of auroras like this one depend on the height and energy of the incoming rays as they collide with atoms of the rarefied gases high above our atmosphere.

flown in aeroplanes, to fill in details of the Earth's gravity field with even greater precision (page 36).

Geomagnetic dating

The crust and mantle are not the only parts of the Earth that are moving, the Earth's core is also being stirred up. It acts as a giant dynamo, generating currents that give rise to the Earth's main magnetic field. This magnetism will turn any particles of iron in a molten rock in a north-south direction. When the rock cools and solidifies the iron is 'frozen' into the direction of the force lines. If that rock is ever tilted or moved geologists can use a special type of compass to detect the movement. This method of geomagnetic dating is now widely used to date rocks on land and on the deep ocean floor (page 29).

▲ Crystals owe their appearance to the chemicals from which they are formed. The shallow rectangles of the barytes (**left**) are quite different from the cluster of radiating crystals of dioptase (**below**).

GEOLOGY IN THE LANDSCAPE

Our Earth has a rough and wrinkled skin. We see its variety all around us, in the contrast of land and sea, of high and low ground. The scenery may include local hills and valleys or soaring mountain peaks. Great mountain ranges such as the Himalayas, Andes and Rockies dominate entire continents. The low plains of Russia and Siberia have little in common with the remote high plateau of Tibet.

Few mountain ranges are peaceful. Many, such as the Andes, are studded with fiery volcanoes. Others, for example the mountains across central Asia, are constantly shaken and fractured by earthquakes. The oceans, too, are varied. Chains of islands are strung across one part while another has only a bare expanse of water. The continents of Africa and South America have facing shorelines that match like pieces of a jigsaw although they are separated by a vast ocean. Below the seas and oceans are submarine landscapes where mountains loftier than the Himalayas rise from the sea bed. The flanks of the land are scored with deep valleys and faced by steep cliffs far below the surface of the waves. How can geologists explain this varied scenery?

The jumbled landscape

Over much of the land soil and vegetation cover the bare rock. In dry or very cold parts of the world sand and gravel drift across the surface or ice and snow hide it all the year round. This is why geologists need to get below the surface of the ground. Sometimes the surface has been exposed to our view, as where the steep bare sides of mountains show clearly how they have been built up. Layers or beds of

◀ In Wyoming, USA, the vertical geological structure of this isolated peak, called the Devil's Tower, is very clear.

▶ In some countries a great mountain range bounds the horizon, as it does here, in Montana, USA.

rock, known as *strata*, may be stacked one on top of another. The strata can be horizontal or tilted; but often they have been folded into loops and wrinkles by the Earth's movements. (You can find granite mountains with no visible structure, such as the Sugarloaf in Rio de Janeiro in Brazil.) Even the smallest islands vary, being made up of sands, coral reefs, smooth rock or stacked layers.

The general contrast between rugged uplands and rolling plains shows up well from the air, but studying the detail needs a closer approach. Every mountain chain consists of lesser units: slopes, peaks and valleys. There may be surprises: extensive caverns beneath green fields are invisible at the surface. Lakes and river courses often pose a puzzle. Why do some valleys hold lakes while others are dry? Why do some rivers abruptly swing into a different course, or even disappear? Why are some mountains steep and jagged, others smoothed into a cone? Geologists can usually answer such questions by looking at the evidence of the rocks and the surrounding scenery.

▲ In central Australia a vast sandy plain is broken by a range of hills called the Olgas.

▼ This regular cone is an active volcano on the island of Honshu in Japan. It is quiet now, and snow lies on its flanks and in the gullies. Other low cones can be seen behind it. Similar but long-dead craters, now covered with soil and vegetation, have been mapped in many countries which nowadays have no active volcanoes.

◄ At low tide the outcrop of rock on which Mont St Michel, north-western France, is built stands isolated among the coastal sands. At high tide it is surrounded by the sea.

▼ A great mass of lava cools relatively slowly and shrinks into regular columns. The Giant's Causeway in Ireland is an example of such structures.

Clues in the rocks

A study of the hardness of the rock may give some answers. If it is loose and *porous*, water runs through it and the valleys are dry. If it is soft clay or a stone without *fissures*, water collects in its hollows. If there is enough rain, lakes or marshes will fill its valleys. Mountains built of smooth hard rock have steep sides. Those built of crumbly material have gentler slopes. But in addition the scenery reflects the climate, both of today and of ages past, and geologists must take this into account.

▼ The scale of this fantastic landscape photograph, taken in Bryce Canyon, Utah, USA, is given by the isolated trees half-hidden in ravines between the serried ranks of pinnacles. Some water must flow through these ravines, at or near ground level, to support these trees. Notice how steep all the pinnacles are. Some are undercut and others have broad flat caps. The pinnacle caps are harder rock than the smooth ridges across the centre of the picture. (Near the top is a plateau of similar rock, not carved into pinnacles.)

THE DIFFERENT TYPES OF ROCK

The rocks that form the Earth vary enormously. They can be divided, or classified, according to their composition, the way that they were formed, and what has happened to them since. Geologists often use complex names for the different rocks, which are understood in many languages, for geology is an international science. These names, like the names used by zoologists and botanists for animals and plants, describe the precise nature of a particular rock.

There are three classes of rock: igneous, sedimentary and metamorphic. Igneous rocks were formed by heat deep within the Earth. (The word igneous comes from the Latin word *ignis*, fire.) They consist purely of mineral crystals. Granite and volcanic lava are igneous rocks. Such rocks are broken up at the surface of the Earth by *erosion*. Then individual grains are dispersed and laid down elsewhere, perhaps mixed with the remains of plants or shells. The whole mass becomes hard-packed and mineral-rich water helps to cement it into a sedimentary rock. Sandstone is a common example of this type of rock. If sedimentary rocks are then very deeply buried, they are again subjected to the great heat of the Earth's mantle. Geologists then call them 'metamorphic' rocks, a name which means 'changed in shape', because the heat changes them into different minerals. Slate and marble are both metamorphic rocks. Everywhere on the Earth's surface you will find broken fragments of rock. Most of these fragments show distinct grains, although you may need to look through a magnifying-glass or hand-lens to see the grains clearly. These grains are identifiable minerals. Each has a definite chemical structure. In some rocks all the grains are the same. But in others, such as granite, you may be able to see white, black and pink grains. Each coloured grain is a clump of crystals. One or more chemical elements combine to form a crystal with atoms packed in a special geometric pattern which also controls the crystal's shape. If the atoms are rearranged into a different pattern, another type of rock is formed. An example of this is the way carbon can become graphite and, under great pressure, may become diamond.

Minerals are formed deep within the Earth's crust and mantle. There, tremendous heat and pressure blend the atoms as if they were in a giant pressure-cooker. If this hot mixture comes to the surface and is rapidly cooled – out of a volcano, for example – then the molten mixture crystallizes quickly. Its tiny grains are packed close together and the rock is smooth and even in colour. In contrast, if the rock stays buried for millions of years it cools very slowly and the crystals have time to grow to a considerable size. Granite usually has large crystals because it is pushed slowly up towards the surface and does not come out fast from a volcano.

From time to time large, perfect crystals are found, usually underground where they have been growing in a fissure with nothing to stop their development.

Not all minerals, however, form as crystals. There are glassy rocks, such as obsidian, and fibrous rocks such as asbestos. The metal mercury – a single element – is even found as liquid drops among the ore in mercury mines.

▼ Tiny drops of liquid mercury can be seen glistening in this quartz rock.

▲ A crystal of red corundum (ruby) reflects the regular arrangment of its atoms.

▶ This table shows the commonest elements in the Earth's crust. They are listed as oxides, that is the element combined with oxygen, because that is how they normally occur. Each has a formula, a kind of scientific abbreviation showing what it is made of. SiO_2 tells the chemist that silica is a molecule consisting of one atom of silicon and two atoms of oxygen.

OXIDE	FORMULA	PER CENT
Silica	SiO_2	59.26
Alumina	Al_2O_3	15.35
Iron	Fe_2O_3 & FeO	6.98
Lime	CaO	5.08
Soda	Na_2O	3.81
Magnesia	Mg O	3.46
Potash	K_2O	3.12
Water	H_2O	1.26
Others		1.68
		100.00

▼ Where convection cells meet, continents are under great stress. This often causes earth movements which cause faults and folds in the rocks. Along the Andean coast of South America the continent is riding over the ocean floor. Where Africa and Europe are coming together, the Alps have been pushed up. Our highest land-mass, the Tibetan plateau, marks the junction of two continents which have piled up, one on the other.

Igneous rocks

Although virtually all the chemical elements are present within the Earth's crust, rocks vary in their mineral content according to where they were formed. Geologists know, for example, that igneous rocks formed under oceans have a different mineral content from those formed under the continents. This is the result of the Earth's convection system (page 12).

The most common igneous rock is basalt. Much of the Earth's upper mantle is a dense type of basalt. We know basalt best as the lava that is expelled through rifts, fissures and volcanoes, beneath the sea and on land.

A basalt plate underlies the Earth's entire surface. It is very thick beneath the oceans, but under the continents it is thinner because their weight pressing on the basalt spreads it out.

Sedimentary rocks

Igneous rocks are transformed into sedimentary rocks in four stages: weathering which breaks rocks down into particles, transport which carries the particles away, deposition which lays the particles down as *sediments* and diagenesis which turns the sediments into rocks. Sedimentary rocks can of course be 'second' or 'third-generation' for they themselves are broken down, carried away and reformed far away, in time and place, from their place of origin.

Weathering is the process that breaks up or disintegrates rocks. The rate of weathering is largely controlled by weather and the land's surface. Plants, rain and frost can break up the strongest rocks, given time. Blocks roll down steep hillsides, fracturing into smaller chips. On level ground, if weathering goes on long enough, the end product will be sands and clays.

The second stage is transportation. Large boulders tumble down from mountains to valleys. They are rolled along by the most powerful rivers. By the time that they reach the sea, perhaps thousands of years later, the rough boulders are polished into smooth pebbles or rounded sand grains. Fine *silts* and clays travel in the slowest streams and rivers. Winds can lift sand and clay and blow it for vast distances, even out to sea. Wind-blown grains are more angular than water-polished ones.

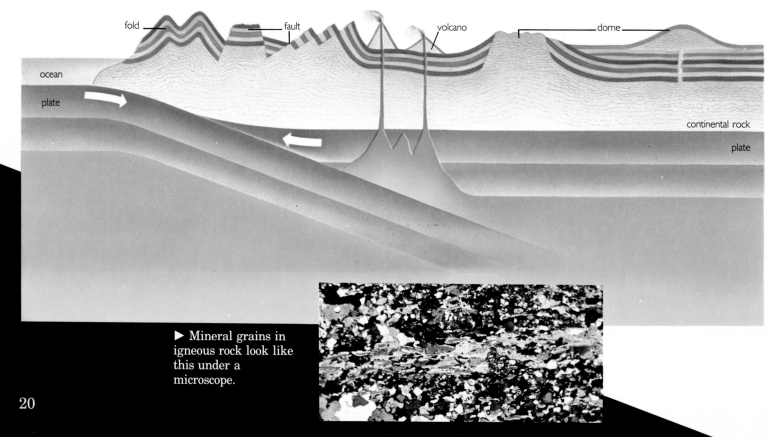

fold   fault   volcano   dome

ocean

plate

continental rock

plate

▶ Mineral grains in igneous rock look like this under a microscope.

20

Deposition is the third stage in the formation of sedimentary rocks. Eventually most sediments find their way into lakes, estuaries or the sea, sinking to the bottom as the current of the stream or river carrying them slows down. By this time most of the rock chunks have been broken or ground down. About three-quarters of the Earth's sedimentary rocks are made of fine silt or clay grains and one quarter are of sand or pebbles.

The fourth and final stage is diagenesis. As more and more sediment accumulates, it presses down on the layers beneath. Water trickles through the sediments, dissolving and redepositing any lime, silica or iron. These act as cement, holding the loose grains together to form solid rock. Under the sea, shells are often mixed in with the arriving sediment. They, too, may be dissolved, to supply much of the lime which cements marine rocks.

The most striking feature of sedimentary rock formed underwater is its stratification. At the surface you can often see alternating strata of hard and soft rock, fine and coarse grains, or varied colours. Each layer represents a change in conditions as the sedimentary rock formed. Perhaps rainfall increased so that flooded rivers carried larger stones than usual. Perhaps stronger winds brought coarse sand in, when before only finer

clay had arrived. Maybe a particular group of plants or animals colonized the area for a while, their plentiful remains mixing with the sediment to alter its colour. Just occasionally sedimentary layers show annual changes from spring flood to summer drought, but generally each stratum represents a far longer time-span than this.

Although most sedimentary rocks form under water they can also form on dry land. Many land-formed sedimentary rocks are ancient desert dunes. Like modern dunes, the sand has built up in a series of sloping layers. Even when the dunes are buried by later sediments they may have to wait for a change of climate before they are flooded and diagenesis can occur. Iron is a common cement for rocks formed on land. It often gives sandstones their rich red or brown appearance. Silica cement, binding pure quartz grains, gives one of the hardest sedimentary rocks known.

Geologists learn most about the Earth's history from the sedimentary rocks which cover over two-thirds of the land. Many were laid down in climates and landscapes quite different from their present surroundings. The majority of fossils are found in sedimentary rocks so they also hold a dateable record of the evolution of life.

Any visit to an estuary or sea shore

▼ Quartz grains are blown by the wind for hundreds or even thousands of kilometres in the world's great deserts. Seen from the air, this desert region displays a large-sized version of the same ripples that form when water washes over sand in a river or on the shore.

today will show you that the top surface of loose sand or mud is sculpted by nature in many patterns. Ripples surge across the flats, leaving a miniature landscape of crests and dips, crossed by sinuous gullies. Ridges of coarse gravel or lines of broken seashells fringe the sea's edge. Birds walk at random across the wet mud, hunting small creatures whose burrows are shown by tiny cones round their entrances. All these features, and many more, can be found by careful searching in quarries or exposed rock surfaces, for they were made in the past, just as they are today. (But remember that there will be no footprints in the older rocks because there were no land animals walking about then.)

The markings on wet sand or mud normally last only a few hours – until the next high tide washes them away. But they will be preserved if a gentle flood of muddy water sweeps across the flats, and fills in all the traces and tracks without spoiling them. Then, if the whole surface is compressed into rock, millions of years later the solid rock may split open at the old land surface, revealing the tracks just as they were formed so many millions of years earlier.

All the rocks mentioned so far are known as 'clastic rocks' as they are made of '*clasts*' or fragments of some other rock bound by a cement. Clasts can range in size from boulders to microscopic clay grains. Very occasionally, however, rocks form purely from a *solution*. Silica is a good example.

Silica can be deposited under water, where it builds up as layers of *chert* filling fissures in limestone. Flint is very similar to chert and is commonly seen as

▲ Marble is a metamorphic rock. It is white when it is pure, but frequently contains other minerals which appear as coloured veins.

▶ When clay grains are compressed into rock they are all parallel, giving it a fine-grained flaky structure. Slate is a clay sediment, well-cemented but easily split. The hardest slates, suitable for buildings, have been thoroughly metamorphosed.

the black nodules in holes and cracks in chalk rock.

When lime-rich water evaporates it leaves behind a deposit known as tufa or travertine. The water bubbling out of hot springs is often rich in lime and other minerals. The crusts that built up around springs in past ages have been preserved in rocks in Africa, New Zealand and America, for example. In caverns, dripping limey water evaporates leaving behind crystals which eventually reach up from the cavern floor or hang down from its roof in the columns we call stalagmites and stalactites. In time, these growths may block up the entire vault and the water will find another outlet to begin the process anew.

Metamorphic rocks
Most of an iceberg is below the surface of the sea. In a similar way the continents have 'roots' into the Earth's mantle beneath the mountain ranges. Deep within these roots the temperature is high enough to melt rock. Here sedimentary rocks (page 20) are so intensely heated that they are turned into metamorphic rocks. With even greater heating the crystals are totally reformed into granite.

When the molten granite rises gradually towards the surface it cools slowly. Crystals begin to develop, spaced well apart so they have room to grow to a fair size before they meet and interlock. Most granite is probably still buried within the roots of high mountains. It may not reach the surface until the mountains themselves are worn away. The great granite 'shields' of Canada and Scandinavia are among the oldest rocks exposed at the Earth's surface today. Scientists believe they were formed between three and four thousand million years ago.

SEDIMENT	ROCK	GRAIN SIZE
Gravel	Conglomerate	> 2mm
Grit		1–2 mm
Sand	Sandstone	60μ–1 mm
Silt		20–60μ
Mud or clay	Shale or mudstone	< 20μ

▲ The grain sizes of sediments and rocks. (The symbol μ is a micron – one-millionth of a metre. Clay grains are less than two-hundredths of a millimetre across.)

THE CHANGING LANDSCAPE

The landscape is changing all the time. You may have seen the sea eating away the bottom of cliffs, until a large area collapses into the sea. Similar events, but on a much larger scale, are going on all over the Earth's surface.

Where *continental plates* meet, the colliding edges may rumple into mountain chains. Elsewhere, the edges of other plates may be sliding down towards the Earth's mantle, dragging with them portions of the crust to vanish for ever into the melting-pot of the Earth's interior. This may sound alarming, but the change is only a few centimetres a year.

Weather at work

All rocks are loosened by temperature changes and the chemicals in water. Crystals expand under the Sun's heat, and shrink as night chills them. Acid groundwater and rainfall trickle through cracks and crevices, eating away the minerals. Rock begins to break up along fissures and strata boundaries. Wind, water and ice sweep away any loose debris so that larger blocks lose their support and fall.

Then, in scientific language, 'all transported material is sorted by the energy capacity of its moving force'. This is just another way of saying that fallen material is only carried, or transported, any further if there is a force strong enough to move it. For example, ice can move larger boulders than fast rivers, but these in turn, can roll and push bigger stones along than the wind, which usually moves only sand or finer particles. However, when the speed of ice, water or

▲ As it winds through the Alps, the Aletsch glacier carries lines of boulders embedded in the ice. Few rivers or streams run fast enough to carry boulders.

▲ Part of the Everglades in Florida, USA. This great area of water and swamp is rich in wildlife. Over a long period of time swamps may fill with the decaying remains of the vegetation that grows in them and so form new land.

▼ A dense stony coral grows as the coral organisms extract lime from seawater to build their protective tubes.

wind drops, the larger materials being carried will also be dropped. They will remain where they were dropped until the speed and strength of ice, water or wind increase again, perhaps in a few hours or next year or perhaps even longer. The mighty boulders lodged in mountain streams may only be moved in times of exceptional flood. But what is important to remember is that nature has plenty of time, and eventually the highest mountain will be reduced and scattered across the face of the land. Ice is the most destructive agent but it only covers a fraction of the Earth at present. Wind and water are everywhere and therefore, though they seem to act more slowly, their relentless action has shaped most of the landscapes we see today or can identify 'frozen' into the rocks.

The work of plants

Wind, water and ice mostly break down and level the land. Rocks may weather and crumble, but where loose material accumulates, soils form. Wherever these are plentiful land can build up. Soils may be scanty in cold dry regions, but up to several metres deep in warmer wetter countries. Almost everywhere there is some form of life. Roots force their way into crevices in the rock, levering it apart. Chemicals liberated by decaying plants accelerate weathering. But humus from decaying plants adds bulk to the soil.

Soils can be buried intact, along with their vegetation, near a river or in a lake, for example, when flooding overwhelms them with fresh mud. Later, a new soil can form on top of the mud. If the land is sinking at the same time, as for example at the Mississippi River delta, the process is repeated until many layers of peat and mud have been stacked up. Many ancient rock strata are capped by thin layers which are the compressed remains of the soils that once covered them.

Changes at sea

Even the sea bed can be built up, in the right conditions. Reef-building corals are colonial animals, living in warm clear sunlit seas. As each coral dies, others grow over it. A little sand may trickle in among the coral skeletons and in time huge banks are constructed, which ring islands or fringe tropical shores. If conditions change and become unfavourable – perhaps the water becomes colder – the entire colony dies. Eventually sediments drift against it and cover it and it is preserved as a feature within the later rock. These stony reefs are known from many parts of Europe and America far distant from the seas where corals live today. Two good examples are the Guadeloupe Mountains of western Texas in the USA and Wenlock Edge in Shropshire, a county on the border of England and Wales.

VOLCANOES

Volcanoes are mountains built of materials brought up through *pipes* or *vents* from the Earth's interior. From prehistoric times to the present day these fiery mountains have aroused fear and wonder. Mount Vulcano, off the toe of Italy, has given its name to Earth's spectacular safety-valves.

Volcanoes occur along the edges of the Earth's *crustal plates*. They are most common around the Pacific, where the ocean floor slides down towards the mantle. Lines of active or ancient volcanoes outline other *plate margins*, in the Mediterranean, and along the African Rift Valley, for example. Volcanic eruptions release the immense pressure caused by cooler lighter crustal rocks being dragged down into the hot dense mantle.

According to the depth of its vent, a volcano may be tapping either crustal or mantle rock. Crustal rock is pale, is rich in silica, has a high melting-point and usually comes out of a volcano as solid cinders or ash. Mantle rock, on the other hand, is dark, contains more iron, has a relatively lower melting-point and emerges from the volcano as molten lava, flowing down the volcano's flanks before solidifying. All volcanoes also emit gases and water vapour. The type of lava and the proportion of gas in it, control the nature of an eruption and the shape of cone that the volcano builds up.

▼ Mount Etna, Europe's highest volcano, is continually active and has over two hundred vents and fissures through which ashes and gas can be expelled. **Inset** Molten lava from a volcano in the Galapagos Islands twists into strange ropes. As it cools and solidifies, the contained gases bubble out.

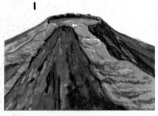

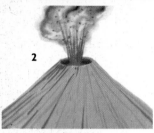

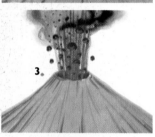

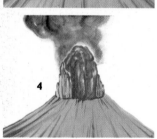

Different types of volcano

Geologists identify five main types of volcanic eruption and have named them after famous examples. They are the Hawaiian, Strombolian, Vulcanian, Plinian and Pelean. Some volcanoes erupt in more than one way; others have changed from one type to another in the course of time.

Hawaiian volcanoes are tapping basalt from far below the crust. Gas bubbles easily from the molten lava which pours down the sides of this type of volcano. The cones have broad, gently-sloping summits. Some have grown to a colossal size as the lava sheets have flooded over them again and again, gradually adding to their height. Mauna Loa, in the Hawaiian Islands, rises 10 km (6 miles) from the sea floor (though only 4 km (2½ miles) show above sea level) and its base is some 100 km (62 miles) across.

Strombolian volcanoes draw their lava from within the crust. It is less fluid and

the gases it contains cannot easily escape. Pressure builds up deep in the vent until it causes an explosion, throwing out glowing cinders and fragments of lava. These spasms occur quite frequently; Stromboli itself normally erupts at least once each hour.

Lava from a Vulcanian eruption is even thicker and high gas pressures are needed to throw it up from the crater. Violent explosions shatter the crater, flinging huge rocks out amongst dense clouds of ash and cinders. When these fall back on the volcano itself they build up a steep cone.

In the Plinian eruption tremendous gas pressure carries a plume of ash high into the atmosphere. The Pelean type throws out more ash than gas, and this rolls down the volcano's side as a glowing avalanche.

In the past volcanoes thought to be extinct have overwhelmed the fields and villages at their feet. The most famous example of this happening was the eruption of Vesuvius in AD 79, when ash and hot mud buried the Roman towns of Pompeii and Herculaneum. In the nineteenth century observatories were built on Vesuvius and Mount Etna, to study their behaviour. Nowadays geologists can usually recognize the small tremors and the pressure build-ups that herald eruptions, though no-one can yet forecast how destructive these will be or whether they will lead to the appearance of a new volcano, as happened with Mount Paricutin in Mexico in the 1940s.

There have always been active volcanoes upon the Earth. Their ancient lava flows and ash cones are found among rocks of every age. Geologists believe that the water of our atmosphere and oceans has come out of volcanoes and the fissures along the plate margins.

▲ Diagrams illustrating four types of volcanic eruption.
1 The Hawaiian type has lava that flows easily over wide areas of land.
2 The Stromboli type explodes at intervals, throwing up ashes, solid 'bombs' of magma. It sometimes pours out streams of lava.
3 The Vulcanian type is still more explosive. It ejects huge rocks, ash and clouds of gas.
4 The Plinian type is the most explosive. It may blow out the interior of the volcano, causing the cone to collapse and leaving an empty shell.

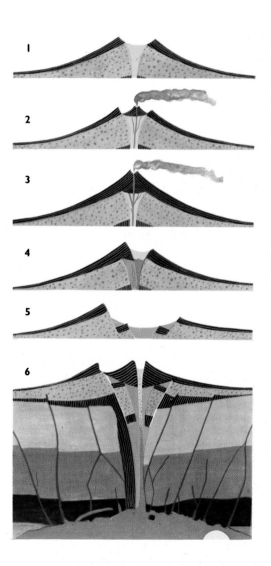

◀ The development of Mount Vesuvius in Italy.
1 The original cone thousands of years ago.
2 A small secondary cone rises in the crater.
3 By the 8th century BC a single cone has formed.

4 The volcano after its eruption in the 8th century.
5 After the eruption in AD79 which destroyed the nearby city of Pompeii.
6 The volcano today, with a vent reaching down to the magma.

FOSSILS

Fossils are the remains of plants and animals, preserved in sedimentary rocks. The word itself comes from Latin and originally meant anything dug from the ground. Later, naturalists used it to describe the stone shells that they saw in many limestones. At first they could not understand how marine animals had found their way into rocks high in the mountains. They did not know how old the Earth was, nor when living organisms had first appeared. Scientists believe that life began some three thousand million years ago. That gives ample time for plants and animals to evolve and change, and for rocks to erode and alter as climates also change.

Fossils are not usually the long-dead creatures but are casts of them. They may form as the buried shells or bones slowly decay. If water rich in minerals crystallizes in the cavities left by the decay at the same time as it cements the sediment surrounding the dead creature, a fossil will form. The fossil is therefore actually a type of rock. Rocks may weather away completely, leaving hard fossils undamaged. In areas of chalk for example, fossil sea-urchins made of flint can often be found.

Fossilization is really a very haphazard process. Most plants and animals are food for something else. Even the parts that may not be eaten, bones, teeth, shells and so on, will be blown or washed away, rolled and battered into fragments. Yet some do survive, to be buried in a quiet lake or sea, covered with a rain of fine sediment. But that sediment must become rock and then be lifted above sea level. Then someone has to come along and discover the fossils. How many millions of creatures must have lived and died for each fossil that we find today!

◄ A shelly limestone, the fossils packed close together and cemented with more lime.

▼ This cast of a sea-urching has been lifted from its surrounding chalk with some spines still intact, showing that it was buried where it died.

► The perfect imprint of a leaf that fell into shallow water some 100 million years ago.

When geologists explored the Grand Canyon of Arizona they found layer after layer of sediments exposed in its steep sides which were over a mile from top to bottom. Clearly this immense thickness represented a very long period of time, but no-one could say how many years had passed since the first beds had formed.

▶ Well-preserved fossils can often be identified very accurately. The fossil (**right**) looks similar to today's clams and oysters. The outlines of the shell plates of the sea-urchin (**far right**) show very clearly. The details help scientists date the rates of evolution.

▼ The Earth has had three ice ages (blue on table). Each one had alternating periods of cold and warm climates (middle column). During the last ice age the ice reached the central USA, Siberia and northern Europe (pale blue on map). Today we are in a warm period. Ice remains only at the South Pole and near the North Pole.

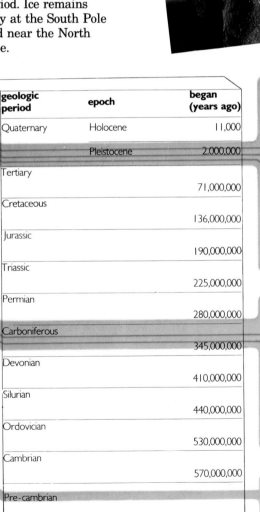

geologic period	epoch	began (years ago)
Quaternary	Holocene	11,000
	Pleistocene	2,000,000
Tertiary		71,000,000
Cretaceous		136,000,000
Jurassic		190,000,000
Triassic		225,000,000
Permian		280,000,000
Carboniferous		345,000,000
Devonian		410,000,000
Silurian		440,000,000
Ordovician		530,000,000
Cambrian		570,000,000
Pre-cambrian		

Pleistocene ice ages	began (years ago)
Wisconsin	(ended 11,000) 125,000
warm age	235,000
Illinoisian	360,000
warm age	670,000
Kansan	780,000
warm age	900,000
Nebraskan	1,150,000

Permo-Carboniferous ice ages	began (years ago)
Permian	280,000,000
warm age	295,000,000
Upper Carboniferous	300,000,000
warm age	310,000,000
Lower Carboniferous	320,000,000

Late Pre-cambrian ice age	
	ended (years ago) 600,000,000
	began (years ago) 900,000,000

North pole

■ present ice-cap limit

□ Pleistocene ice-cap limit

Relative dating

Strata such as those in the Grand Canyon can be relatively dated. That is, we know that when sediments are laid down, the oldest are at the bottom and the youngest are at the top. But sometimes mountain building and earthquakes fold and tilt strata so that their original order cannot be seen. Geologists then find fossils useful guides to the original stratification of the rocks. Fossils can also show that rocks a long way apart are of the same age. For instance, if fossils of a particular species are found in Africa and in South America we know that both sediments are the same age, although they may look very different.

Radiometric dating

Such methods of relative dating are most useful if we know the true age of at least some of the layers. Geologists can now date igneous rocks by measuring their radioactivity. The technique is called radiometric dating. It is based on the fact that some elements in the Earth's crust, such as uranium, disintegrate naturally into a lighter element. Over millions of years one atom after another emits a particle from its nucleus and is transformed into another, *stable element*. Radioactive uranium is transformed into stable lead. This process is extremely slow, but regular. Scientists analyze samples of rock and the ratio of radioactive to stable atoms reveals its age.

Radiometric dating is a complex procedure and must be used with caution. Although radioactivity was discovered at the end of the last century, the apparatus that enables scientists to date rocks has only been available since about 1950. Even so, enough samples of lava, ash and other igneous rocks have been analyzed to provide a framework for dating the Earth's history. Besides the remote past, some more recent volcanic ash layers have been dated. In sediments among these ash layers fossils of our own ancestors have been found. We are beginning to have dates for the evolution of early man and of *Homo sapiens*, our own species.

Dating the sea floor

As more and more iron-rich basalts were dated, geologists made a surprising discovery: the Earth's magnetic field had reversed many times in the distant past, with north and south poles changing places. This discovery has been particularly useful in helping scientists discover how the sea floor has spread and changed over many millions of years.

As basalt wells up from the mid-ocean rifts its iron particles align themselves with the Earth's magnetic field. As the basalt cools and solidifies this alignment is 'frozen' into the rock. As the sea floor spreads, strips of basalt, their internal compass needles pointing alternately north and south are carried away on each side of the rift. A ship towing a *magnetometer*, a sensitive magnetic recorder, across the ocean at right angles to the rift can record these patterns. The patterns are then checked against the dated basalts to discover when the basalt strip was formed and how far each strip has travelled since then. The rates of movement vary. The sea floor is spreading at about 1cm (½in) per year near Iceland, but at 9cm (3½in) per year at the mid-Pacific ridge.

As a result of these measurements, scientists now know that the ocean floors are the youngest part of the Earth's crust, probably no more than 200 million years old. This makes them much younger than the continents. The fact that the sea floor is quite 'young' also supports the theory that as the sea floor has spread in one place, so it has disappeared, perhaps under the edge of a continent, somewhere else.

▼ As iron-rich basalt emerges at mid-ocean rifts the iron particles align themselves with the Earth's magnetic field. The basalt is carried away on its moving plate, each strip with its own magnetic 'signature'. By comparing these 'signatures' with known and dated changes in the Earth's magnetism, we know when the basalt emerged, and the rate of its movement. In this way we can date events such as the separation of America from Europe and Africa.

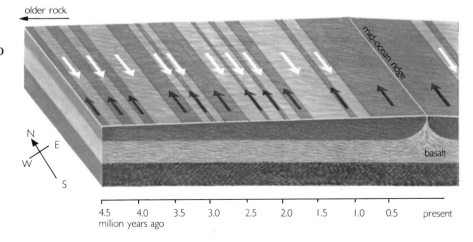

older rock

N
E
W
S

mid-ocean ridge

basalt

| 4.5 | 4.0 | 3.5 | 3.0 | 2.5 | 2.0 | 1.5 | 1.0 | 0.5 | present |

million years ago

OUR MINERAL CIVILIZATION

THE FIRST USERS OF MINERALS

Archaeologists name the past stages of our culture the Stone, Bronze and Iron Ages, after the different materials that men learnt to use, each one a little more complex than the one before. The Stone Age began a million or more years ago, long before *Homo sapiens* appeared on Earth. Our ape-like ancestors discovered that some kinds of stone could be chipped into tools. Later, they found out how to use bronze and then iron. We still rely on plant and animal products, for timber, paper, fibres and food, but now the mineral kingdom supplies most of our needs. In fact, over the past 5,000 years, our use of minerals has increased so much that we are in danger of exhausting the reserves.

These reserves are the so-called economic mineral deposits. The entire crust of the Earth is made of minerals, but some are concentrated into limited parts of the world. The economic deposits include metal *ores*, such as iron, non-metallic minerals such as diamonds, and *hydrocarbons* such as oil. The first two have long been in demand and supplies have been available at a price consumers were willing to pay. Hydrocarbons, on the other hand, only become important in the last hundred years and their development is expensive.

◀ Hematite ores of iron are so distinctive that they soon attracted attention. Nowadays other more plentiful, but less attractive, ores are worked for iron.

▶ Some metals and other useful minerals occur naturally in the rocks and so were among the first to be used. This native silver comes from the Harz mountains, Germany.

Even in prehistoric times minerals and some metals were collected. Copper, gold and silver could all be found in their natural state at the surface of the ground. Gems and semi-precious stones were highly valued and traded far and wide from their source. From the days when men first painted cave walls, coloured minerals have been ground up to make pigments and dyes.

We will probably never know who first discovered that some lumps of earthy-looking rock were actually ores that could be converted into a material tougher than stone. If these lumps of rock were heated they yielded a soft or molten metal that could be hammered or cast into shape. The metal could also be alloyed – mixed with another metal – and so made even stronger. Bronze is an *alloy* of copper and tin. But these two metals seldom occur together naturally, so the Bronze Age peoples were clearly knowledgeable and skilful. They also began to search out the ores of tin and copper, learning to

recognize where they might be found. They built furnaces to *smelt* and refine the ores. Gradually other metals were discovered together with the ways to work them. For example, iron ore is more plentiful than copper or tin ores, but is smelted at higher temperatures.

▲ Contemporary picture showing miners digging and drying alum (for use in the dyeing industry) in the early 19th century.

The first mines

With iron tools it becomes easier to dig away soil and rock to search for other minerals. Shallow pits were extended into the hillsides or sunk into the ground. Nations found themselves richer or poorer, according to their share of the Earth's mineral resources. Trading networks grew up, linking ores to the refineries where fuel was abundant. As the exploitation of the Earth beneath us gathered pace, the engineers of the eighteenth and nineteenth centuries introduced machines to make the work of extracting the minerals more efficient. Mechanical pumps and hoists were installed. Ventilation took air down to the hot damp pits. Drills and dynamite reduced the amount of backbreaking labour spent on each ton of ore sent to the surface.

The demand continues to increase and the search for new supplies has led geologists to intensify their studies of the hidden rocks. Most of the recent important discoveries about the Earth and its past history have come from prospecting for more minerals to satisfy this ever-increasing demand.

IRON, ALUMINIUM AND COPPER

Most metals do not occur naturally in a pure state, so the ore must be smelted to extract the metal. Iron, aluminium and most copper have to be smelted from their ores.

Iron

Iron ore, in particular, is often full of impurities and may contain only about 15% iron. In some parts of the world, however, there are natural concentrations of sedimentary ores containing up to 60% iron. The rich Brazilian and Australian iron ores are examples of this.

Aluminium

Aluminium ore is called bauxite. It is generally a recent formation, and is found close to the surface in tropical latitudes. Bauxite forms when aluminium-rich rocks are intensely weathered. In a hot climate, with abundant rainfall and good drainage, the rocks will be partially dissolved. The aluminium compounds remain and gradually accumulate in thick layers as the parent rock weathers away.

Copper

Primary copper *ores* have less than 1% copper spread through otherwise barren rock. Such ores form in regions of volcanic activity. They are mined close to recently formed volcanic arcs (page 34), in New Guinea and the Philippines. Copper from ancient volcanic regions has now, like some iron ores, been deposited as rich sediments. There is, for example, up to 8% copper in the African Copper Belt deposits.

Ores where copper is combined with sulphur yield only 5% copper, often mixed with zinc and other metals. Such ore bodies are generally smaller than other copper deposits but they are still worth mining. The Kidd Creek open cast mine, in Ontario, Canada, has been worked down to some 250 m (820 ft). As the ore body continues down for at least another kilometre, it will be worked as a deep mine, with the ore being brought up through shafts and tunnels.

▼ Australia is developing its mines and refineries to supply markets in nearby Japan and East Asia. Here iron ore is dug from the appropriately named Iron Knob Mine in South Australia.

MINERALS FROM A MOVING CRUST

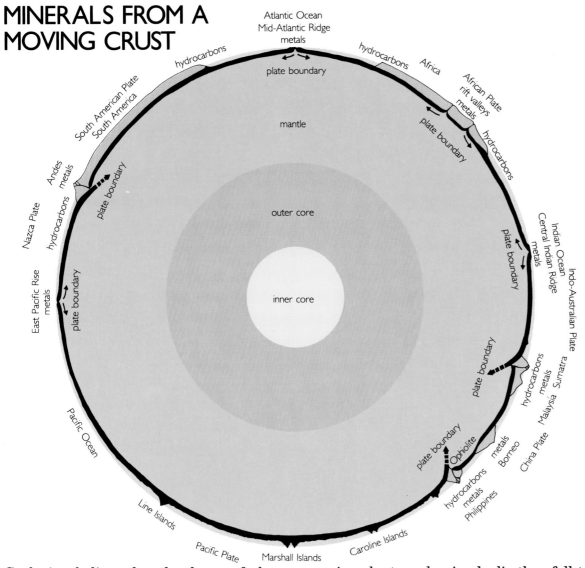

Atlantic Ocean
Mid-Atlantic Ridge
metals

hydrocarbons

plate boundary

Africa

hydrocarbons

African Plate
rift valleys
metals

plate boundary

hydrocarbons

South American Plate
South America

mantle

Andes
metals

plate boundary

Nazca Plate

hydrocarbons

outer core

plate boundary

Indian Ocean
Central Indian Ridge
metals

East Pacific Rise
metals

plate boundary

inner core

Indo-Australian Plate

Pacific Ocean

plate boundary

hydrocarbons
metals
Sumatra

Malaysia

plate boundary

Ophiolite

metals
Borneo

China Plate

hydrocarbons
metals
Philippines

Line Islands

Pacific Plate

Marshall Islands

Caroline Islands

◀ Diagrammatic section through the Earth at the equator to show the Earth's crustal plates, the boundaries of the plates and where energy (hydrocarbons) and mineral resources are found. The arrows show the direction in which the plates are moving.

Geologists believe that the theory of *plate tectonics* can explain why metal ores, hydrocarbons, salt and other important minerals are found only in certain parts of the world. According to this theory the Earth's crust consists of huge moving plates whose boundaries are the *seismic belts*. Heat from the upper mantle escapes along these boundaries. It is here, if conditions are right, that the various important minerals are formed and concentrated.

Minerals where continents split

As the continents move across the globe, ocean basins open and close at different latitudes. The Red Sea is an example of a new ocean basin. It is forming as the African continent is torn apart from the valley of the Dead Sea through the Rift Valley system to the Indian Ocean. The following sequence of events takes place. The continent splits or rifts. First lakes and then an enclosed sea form. When

marine plants and animals die they fall to the sea-bed. But as there is not enough oxygen to make them rot quickly, they are buried by sediment washed off the nearby continent. In a hot dry climate, like that of the Red Sea area, the water in the enclosed sea evaporates faster than fresh water flows in. As a result the sea's *salinity* rises until salt is deposited. This rock salt can form beds several kilometres thick.

As the continent is still splitting, the sea expands until the ocean can wash freely in and out of it. Then the environment changes. The water is fresher, so no more salt is deposited. It has more oxygen, so dead organisms decay quickly. But sediment continues to arrive from the land, building up the *continental shelves*. Under the pressure of this load, and heated by the movements of the mantle below, the undecayed organic matter is converted into gas and oil.

As the shelf sediments are compressed

and folded, salt, gas and oil which are quite light migrate upwards through the sediments. The salt is often squeezed into domes or pillars. When oil and gas are trapped under a layer of non-porous rocks, such as salt domes, clay or shale, natural reservoirs form and it is here that the world's reserves of gas and oil are found. If they are not trapped the gas evaporates into the air when it reaches the surface, and the oil seeps away in natural slicks.

As the land mass continues to divide and the new ocean gets larger, seawater and water from deep within the crust penetrate fissures along the rifting floor. The water becomes hot enough to dissolve any metals in the basalt. This hot fluid, accompanied by hot gases, gushes up from vents along the plate boundary. Here the dissolved metals are deposited as crusts and sediments that may be several metres thick. They are rich in metals such as iron, manganese, copper and zinc.

As ocean basins open and close there is more or less space for the water on the surface of the globe. At the present time some water is locked up in ice-sheets. But when there are few deep oceans, and the global climate is mild, the sea-level will be higher than it is now. Water will flood over low plains, creating broad shallow seas. During the Permian period, which lasted from 280 to 225 million years ago, broad seas washed over the western and south-eastern parts of the north American continent. A gulf of the polar ocean stretched far into northern Europe, extending at times even into Siberia. Many of our useful chemicals, including salt, potash and gypsum, were laid down by these inland seas.

Minerals where continents collide

When moving plates meet, one may slide beneath the other (if this doesn't happen, they will crumple upwards to form mountain ranges like the Himalayas). Moving down towards the mantle, the lighter rock creates a disturbance as it mixes with the basalt. This disturbance throws up a line of volcanoes parallel to the plate boundary. Offshore volcanic arcs are forming in this way along the Pacific coast of south-east Asia and Alaska. On the American coast the volcanic arc is inland, among the peaks of the still-rising Andes and Cascades. Oil and salt often occur in seas separated from the ocean by such off-shore volcanic arcs.

Metals, as well as oil, occur where plate margins meet. They form in a similar way to those minerals that occur where the continents are rifting apart. The plunging oceanic crust of one plate carries water down from the sea. This water is heated, dissolves metals from the surrounding basalt, and is swept back through fissures to the surface where it deposits the minerals as crusts and sediments.

▼ These diagrams show how mineral resources probably developed in the southern part of the Atlantic Ocean. The ancient supercontinent of Pangaea divides in two (below). As the South American landmass moves west (middle), an ocean trench forms. Thick layers of rock salt, organic matter and metallic minerals are deposited in the sea. As South America continues to move west (bottom), pressure with a crustal plate causes the formation of the Andes. Minerals that melted as the Pacific Plate went under South America ascend to become mineral deposits in the Andes.

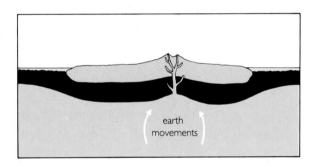

earth movements

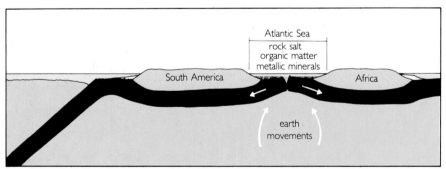

Atlantic Sea
rock salt
organic matter
metallic minerals

South America

Africa

earth movements

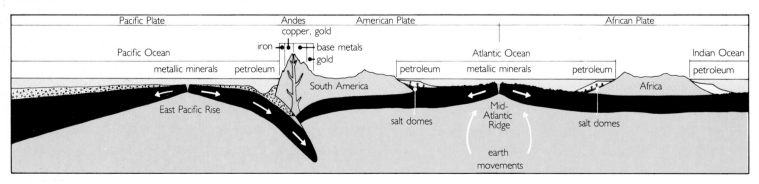

Pacific Plate	Andes	American Plate		African Plate
	copper, gold			
	iron — base metals			
Pacific Ocean	gold		Atlantic Ocean	Indian Ocean
metallic minerals	petroleum	petroleum	metallic minerals	petroleum	petroleum
		South America		Africa

East Pacific Rise

salt domes

Mid-Atlantic Ridge

salt domes

earth movements

THE SEARCH FOR MINERALS TODAY

Minerals, particularly metal ores, seldom occur singly. Groups of ores, such as copper and nickel, or lead, zinc and silver, occupy the same *vein*. Some non-metals, such as salt and sulphur, are also found close together. The *prospector* knows, when he finds one of these minerals, that the others are probably there too.

In the past, most minerals were identified from outcrops on the Earth's surface. Patient observation of the presence (or absence) of different plants, the appearance of the soil, a change in rock type or analysis of chemicals in streams all enabled prospectors to find the minerals they wanted. Most of these

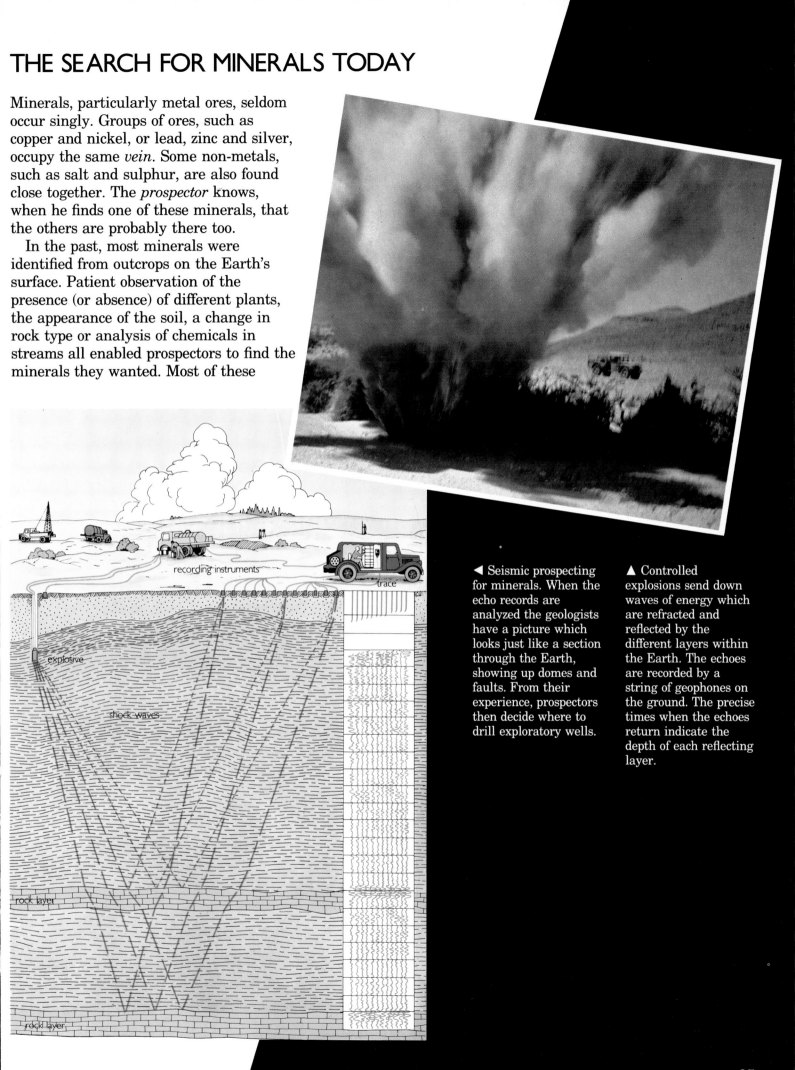

recording instruments

trace

explosive

shock waves

rock layer

rock layer

◄ Seismic prospecting for minerals. When the echo records are analyzed the geologists have a picture which looks just like a section through the Earth, showing up domes and faults. From their experience, prospectors then decide where to drill exploratory wells.

▲ Controlled explosions send down waves of energy which are refracted and reflected by the different layers within the Earth. The echoes are recorded by a string of geophones on the ground. The precise times when the echoes return indicate the depth of each reflecting layer.

► However promising the geology of an area, only test drilling can confirm if there is oil there. These drill cores show black patches of oil filling cracks in limestone.

► If surface rocks are mapped to show their angles of dip and slope we get clues to the depth and location of hidden mineral beds which outcrop at the surface. In this map, the coal seams at the top right-hand corner dip to the right. It would be useless to prospect for coal under the mountains in the middle of the map.

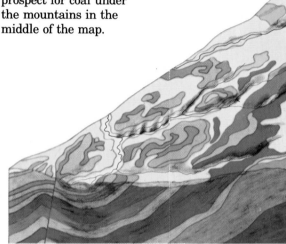

▼ Australia is rich in mineral resources. Other countries may be less fortunate; their valuable minerals have been eroded away in the past. But nowadays even deep deposits can be found and worked.

minerals

△	oil
▲	gas
●	coal
Cu	copper
Pb	lead
Zn	zinc
Al	aluminium
Fe	iron
Au	gold
Ni	nickel
U	uranium

vegetation

	cultivated land
	pasture
	forest
	savanna
	desert

accessible deposits are now used up. Today's prospectors have to use more elaborate methods to detect ore bodies far below the surface.

Ore and mineral bodies are quite distinct from the rock surrounding them. They may be more magnetic, or conduct electricity better. They may be lighter or heavier, size for size. They may be denser or more porous. Most modern methods of prospecting therefore involve measuring the Earth's magnetic field, its gravity or its density, in a particular region to find these differences. Prospecting in this way with a magnetometer, an aerial survey discovered a huge iron-ore body 450 m (1,500 ft) below densely-settled areas of Pennsylvania. Nickel deposits in Manitoba were found by the same method.

Gravimeters are used to find useful minerals which are lighter or heavier than the surrounding rock. Where they occur in quantity they alter the local force of gravity, and the gravimeter detects the tiny changes they cause.

Radioactive rocks such as uranium ores emit radiation. They can be located with a Geiger counter. But seismic techniques (see pictures page 35) offer the best

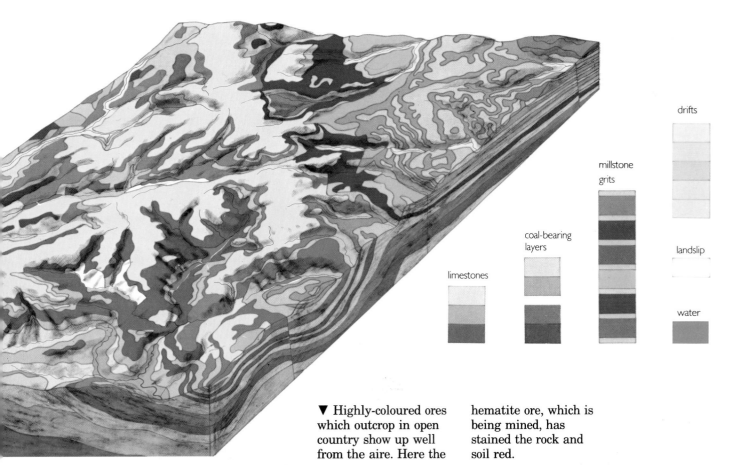

drifts

millstone
grits

coal-bearing
layers

landslip

limestones

water

▼ Highly-coloured ores which outcrop in open country show up well from the aire. Here the hematite ore, which is being mined, has stained the rock and soil red.

picture of underground structures. They are increasingly used on land and offshore to find the best sites for exploratory drilling.

▶ A lump of hematite as excavated from the ground and before it is processed to become iron.

COLLECTING MINERALS FROM THE EARTH'S SURFACE

The majority of the Earth's resources are still taken at or near the surface. Some occur there naturally. Around the Mediterranean, for example, salt is still taken from the sea by evaporation. Soda and potash are collected from the Dead Sea. Sulphur is still gathered near volcanic vents in Sicily.

Some minerals are so widely distributed through the crust that there is enough available at the surface without mining. Examples are building stone, sand and gravel, limestone for cement and clay for bricks and tiles. Most countries have adequate supplies of these essential building materials and they are all fairly cheap to obtain.

Although most of the world's gold and gemstones are mined, occasionally they are still found on the surface. Originally, the gemstones and gold formed in veins deep within the crust. The rocks holding these veins were later folded up into mountain ranges. Earthquakes shattered the rock strata and the weathered material was carried down to the rivers. As the lighter sands and mud were washed downstream the hard gemstones and heavy nuggets of gold remained.

▶ These crystals are all diamonds. The clear ones are pure carbon, and the most valuable as gems. Impurities have coloured the others. Diamonds are the hardest natural material known. Any rejected by jewellers are used as industrial abrasives for cutting, drilling and polishing other stones.

◀ Stacking blocks of sulphur at a sulphur mine in Sicily. This mineral is always found close to present and past volcanoes. In Sicily it is collected at the surface, as it was in ancient times, and is also mined from below ground.

◀ Honey-coloured limestone columns on the Acropolis in Athens, Greece, are suffering badly from erosion. Like natural rock in the field they will eventually crumble into dust unless scientists can find a way to protect them or to change them into a harder stone.

▲ Outcropping granite is being sawn into blocks for building stone in this quarry in northern Italy.

OPEN-CAST MINING

The valuable minerals, iron, aluminium and copper, on which so much of modern life depends, are extracted and processed by huge international mining companies. Where the ore lies close to the surface the top layers of vegetation and soil are stripped off and the ore dug out. This is open-cast or strip mining.

If the ore deposit is deeper than about 300 m (1,000 ft) a deep mine (page 42) is made.

Mining can be very destructive of land. Ideally, mines are located in deserts or polar regions, or in relatively uninhabited parts. But the demand for these metals is so great that they must be mined almost wherever they are found.

◄ The world's largest open-cast copper mine is at Bingham Canyon, Utah. To keep up with demand, ores containing less and less metal are being worked in some countries.

▲ Sweden has supplied Europe with iron ore for many centuries. Today, with advanced technology, entire mountains of iron ore in the Arctic north are quarried away and processed.

◀ Bauxite is a residue of many tropical soils. It is normally found near the surface and is taken from open-cast mines. So much energy is needed to obtain aluminium from the bauxite ore that refineries are usually built where cheap hydroelectric power is available – often a long way from the bauxite mines.

DEEP MINES

▼ This illustration shows four types of mine. Open-cast mines are suitable when the deposits are close to the surface. In mountaineous country drift and slope mines may be used to reach the deposits without removing overlying rocks. Shaft mines run directly from the surface to the deposit below.

Deep mining is a costly business. Mine shafts and tunnels have to be dug, then lit, ventilated and pumped dry. Men and machines have to be taken to and from the working surface. An enormous volume of useless rock may have to be taken out of the workings along with the mineral being mined. The cost of all this has to be balanced against the value of the end-product. This means that deep mines must have better-grade ore, or yield a rare or costly product.

Deep mines generally stop at depths of about 1800 m (6,000 ft) depth. A few go down to 3 km (10,000 ft). At depths like this, reached in some diamond mines, the heat is intense. About three tons of gem-quality diamonds are mined each year, most of it from South Africa. Geologists believe that the carbon from which diamonds form was compressed under the enormous pressures of the Earth's upper mantle. In South Africa, pipes of this mantle rock have been injected up into the crust. The miners excavate these pipes, following them down as far as modern techniques permit.

If ore bodies are of loose rock, or can be shattered by underground explosions, it is sometimes possible to leach the ore. To do this water is sent down to dissolve the ore. The natural heat and pressure below ground encourage this process, and the enriched solution is pumped up to the surface to extract the minerals. There is a limit to the depth that can be leached in this way. Below about 6 km (3¾ miles) the pressure is so great that it closes all the rock pores and water cannot wash out the ore.

A similar method is used in Texas and the Gulf area, where there are deposits of sulphur between 150 m (500 ft) and 1 km (3,300 ft) below the surface. The sulphur is melted by injecting hot water and then a froth of water, air and sulphur is pumped back to the surface.

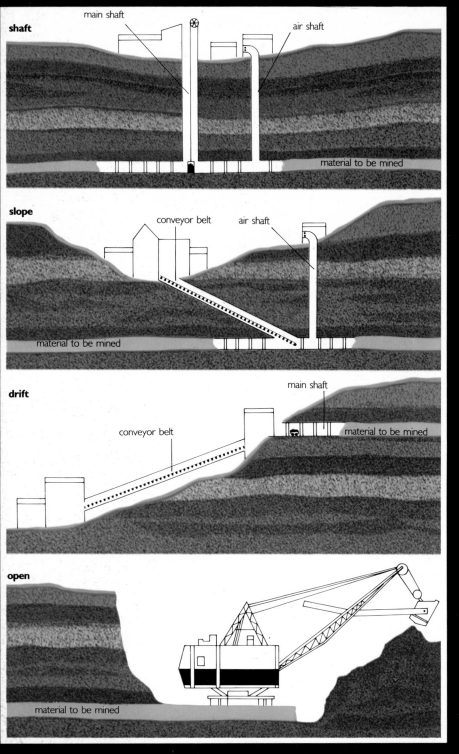

shaft — main shaft — air shaft — material to be mined

slope — conveyor belt — air shaft — material to be mined

drift — main shaft — conveyor belt — material to be mined

open — material to be mined

▲ Corundum, a compound of aluminium and oxygen, is one of the harder minerals and is used as an abrasive. Ruby and sapphire are forms of corundum stained by impurities.

▲ A great volume of unwanted rock may have to be shifted out of the mine to give the miners working-space, especially if the ore-bearing vein is thin and narrow.

◄ Inflammable gases, dust, and the constant risk of flooding or tunnel collapse make the underground mining of any mineral a hazardous and costly business. Very deep mines are uncomfortably hot due to the increasing temperature below ground.

43

MINERALS FROM SEAS AND RIVERS

▼ Dredging for tin ore in Malaya. In this region the tin-bearing rocks have been eroded, leaving behind the heavy grains of cassiterite mixed with river gravels.

Inset
Dredgers scoop sand from shallow water and carry it away for screening for heavy minerals. In deeper water sand can be lifted by huge suction-pumps, directly into lighters which ferry it back to land. As well as being used for building, sand is the main constituent of glass.

Mineral extraction is not limited to dry land. It is often worth searching under water. Offshore dredging in river estuaries, lakes and shallow seas yields most of the world's sands and gravels which have already been sorted by underwater currents. The stronger the current, the larger the pebble that it can sweep away. For example, in the southern North Sea, where much commercial dredging is done, banks of sand accumulate in slack water while gravel is found along channels where strong currents clean out any mud or sand.

Metals such as tin are now regularly worked offshore, for although tin has been used since ancient times, it has always been scarce. It is never found as a *native metal* but must be refined from its ore, cassiterite. In a few parts of the Tin Belt, stretching from China across south-east Asia, it has been washed out of its source rock and is found concentrated in gravels offshore. Elsewhere it is mined, as in the Andes in Bolivia. In the Kidd Creek mine in Canada, it is found with copper ore.

Certain useful chemical elements, for example iodine and bromine, can be extracted directly from seawater. These chemicals also occur on land, having evaporated from long-vanished seas. There are good deposits from one such sea that covered much of central Europe. But today most of the world's supply of iodine and bromine come from enriched brine wells, such as those in California, where ancient marine deposits have concentrated these elements.

buckets

ENERGY FOR THE WORLD

◄ How coal forms: the process from forest trees to coal seam. Coal crumbles easily and has been eroded from many of the seams it once filled.

Small quantities of coal and oil have always occurred at the Earth's surface, but only since the nineteenth century have they been used so much. Besides fuel, coal and oil provide us with many things, including dyes, drugs and plastics.

Hydrocarbons are formed from decomposed plants and animals. This organic material consists of fatty *molecules* plus carbon. When this material decays under water without oxygen bacteria digest the organic remains, breaking or *cracking* the large molecules into smaller units plus kerogen, which is basically carbon. The process continues as the sediment is buried and heated by compression. Two factors determine whether the end-product will be coal, oil or gas. The first factor is the original ingredient: coal forms from plants and oil and gas from tiny animal life of the sea. The second factor is the depth to which the sediment is carried: gas occurs at the greatest depths while coal is nearest the surface with oil in between.

Coal

Coal is accumulating today in the Mississippi delta where plant debris and sand build up at such a rate that the delta is sinking under their weight. This carries the debris down to depths where compression and high temperatures will turn it into coal. Coal has only formed of course since land plants evolved. Most comes from rocks of the Carboniferous period (345 to 280 million years ago) and the Permian period (280 to 225 million years ago). Softer coal, which has not

▼ With modern
technology it is now
possible to drill for oil
and gas in the open
sea. Huge rigs like this
one are anchored over
the well-head far below
on the sea floor.

► Oil and gas often
emerge from the same
well. Here, in south-
east Iraq, the oil is
being piped to a
terminal while gas is
flared off as a waste
product. In other fields
the wells yield mainly
gas, and the oil is not
worth collecting.

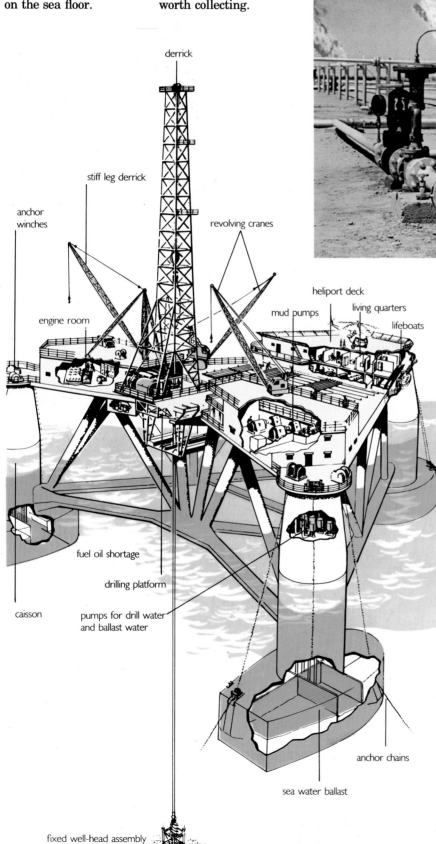

derrick

stiff leg derrick

anchor
winches

revolving cranes

heliport deck

mud pumps

living quarters

lifeboats

engine room

fuel oil shortage

drilling platform

caisson

pumps for drill water
and ballast water

anchor chains

sea water ballast

fixed well-head assembly

been so deeply buried, is called lignite, or
brown coal. It forms from peat which is
being laid down in many parts of the
world today. If this peat is in deltas
enough material will probably collect on
top of it to compress it into coal.

Oil

Oil forms at depths of around 2 km (6,500
ft). If the sediment sinks deeper, the
molecules are broken down still further.
At a depth of 3 km (10,000 ft) they have
become methane gas. This conversion
from complex organisms to simple
molecules takes at least one million years,
and generally longer.

Oil and gas are lighter than kerogen
and the surrounding sediment. They
diffuse through pores and fissures, moving
slowly towards the surface. If the small
amount of kerogen that is left continues
to sink it may ultimately become
graphite, a pure form of carbon.
Hydrocarbons which reach the surface
will seep out or evaporate. If they are
trapped by a layer of impervious rock
they will form a natural reservoir (page
34). But even then they may be lost if
that cap rock is fractured by later earth
movements. This explains why so many
promising sites turn out to be dry wells.

▼ Steam and water gushes from an active geyser in Yellowstone Park in the Rockies, USA.

► An oasis is a natural or man-made depression that breaks through the hard capping rock and allows water from the aquifer to fill a pool where plants can grow and animals can drink.

Water is not a mineral, but it is essential to life. By tapping underground water humans can settle even in dry lands where there is not enough rainfall for crops and livestock. But why is water present under desert sand where today it seldom rains?

All moisture on this planet is part of what geologists term the Hydrological Cycle. Water from the land and ocean surface evaporates into vapour. It rises into the air, condenses into clouds, and eventually falls back to the land or ocean as rain. Some of this rain runs directly into rivers. More seeps through loose sediments or fissures in the rocks, sinking till it meets some watertight (impervious) layer, such as clay or solid rock.

The level to which the water sinks is called the water-table. The areas now occupied by the world's major deserts once had much wetter climates. They still have water trapped below ground on an impervious layer. In the Sahara Desert rain that fell millions of years ago is held in beds of loose sandstone not far below the surface. The sandstone slopes down towards the north, so this water is gradually flowing towards the Mediterranean Sea. Wells are sunk to tap this layer of water, but it is now being used up so fast that there is not enough rain to replace it. Each year, as the water-table falls, the wells have to be deepened to reach fresh supplies.

Groundwater that has penetrated great depths in deeply-folded rocks becomes heated. If it can reach the surface quickly it may emerge as a hot spring or an explosive geyser, forced out by steam pressure. Such waters are often highly mineralized. The water evaporates, leaving behind terraces and waterfalls of the minerals, often brightly coloured. The hot springs and geysers of Iceland, the North Island of New Zealand, and Yellowstone Park in the USA are famous examples. The warm spa waters of Europe are less spectacular but have been valued for their medicinal properties since Roman times.

Mystery room
Lower cave
New Mexico room
Big room
0   200   400 metres
entrance (from surface)
left hand tunnel
Bat cave
Lake of the clouds

◄ A map of the Carlsberg Caverns in New Mexico, USA. The system extends for about 6.5 km (4 miles). The largest caverns could each hold ten football fields and some are as much as 300 m (980 ft) under the ground. It has taken 60 million years for the caverns to reach their present size.

► Water trickling through soil and decaying vegetation becomes slightly acid. It can then erode limestones very effectively, sometimes into fantastic cave systems. The rainfall percolates through porous levels to collect in some impervious basin. Outflow from the basin may be many kilometres distant from the water's original source. These are the Punkevni caves in Czechoslovakia.

4 GEOLOGY AND THE FUTURE

▲ The original lines of mountain folds can be picked out in this photograph of the Pindus mountains in western Greece, although erosion has carved deeply into them.

SATELLITE MAPPING

Geology is one of the youngest sciences. Before the nineteenth century few people ventured into the high mountains or barren lands where rocks are seen at their best. The excavation of canals and railways in Europe brought surveyors face to face with the strata exposed in the

cuttings and the first geological map was drawn up by William Smith in 1815. As the western parts of America were opened up, geologists joined surveyors in the field, and found the Earth's history laid out before them on a grand scale. A century of intensive field work followed, with geologists travelling to all parts of the world. They were searching for

50

minerals to replace the exhausted supplies of Europe and, especially during the 1930s, for petroleum. They were studying volcanoes and earthquakes, to see if these could be understood and forecast. They were also seeking fossils to build up a picture of the Earth's evolutionary history.

Where men and vehicles could not penetrate, aircraft took over, carrying cameras and sensitive instruments. In our own time these cameras and instruments have been transferred to satellites. The data that is sent back justifies the cost of this work, for the maps and all the measurements over land and sea have increased our knowledge of the Earth.

Satellites carry cameras that not only

▲ A satellite photograph of the Gulf of Genoa in north-western Italy. The River Po winds across the top of the picture. Smaller rivers have cut valleys in the foothills of the Apennine mountains as they flow north into the Po.

take conventional photographs but also
pictures at infra-red and radar
wavelengths. Infra-red reveals hot-spots
in the crust where basalt or granite has
been thrust upwards from the mantle.
Radar penetrates soil and vegetation to
photograph the hidden bedrock.

Geophysical measurements made from
satellites give us a far better
understanding of the whole Earth, in
three dimensions, than could ever be
achieved from the ground. These
measurements pick up variations in the
Earth's gravity and density, so revealing
the composition and depths of the crustal
plates. They also tell us something about
the hidden surfaces of mantle and core.
Geologists share with astronomers and
physicists an interest in the
magnetosphere – the region high above
the atmosphere where the Sun's magnetic
field envelops our planet. Data collected
from the magnetosphere may provide
information about the origins and
maintenance of the Earth's own magnetic
field.

GEOTHERMAL HEAT

In countries such as Iceland natural hot springs are used to heat homes and greenhouses. At Larderello near Pisa in northern Italy, steam from below the ground has been providing electricity since 1904. Now satellite infra-red photographs are showing up other regions which could be used in a similar way. In Cornwall, England, for example, there is a scheme to drill into one such hot-spot. First, several holes will be drilled to depths of 2–3 km (6,500–10,000 ft). Then explosive charges will be used to fracture the rock between the shafts. Finally, water will be pumped down to be heated before returning to the surface under pressure from the pumping down of more water.

DEEP SEA AND ANTARCTIC MINERALS

Huge mineral resources lie under the deep seas and the ice-cap of Antarctica. Manganese nodules were discovered over a century ago. These nodules also contain copper, cobalt and nickel. They cover wide areas of the deep ocean floors, and till now have been beyond the reach of commercial dredgers. As land-based resources become scarce, mining companies are looking for ways to exploit them at a profit. There are, as we now know, spreads of concentrated metal ores along many of the undersea plate margins. If the technology can be developed, both these new sources will become available. But the open oceans and the Antarctic continent lie outside national territorial limits. Before any company can take these minerals, the nations of the world will have to agree on their ownership.

▲ Small manned submersibles place apparatus on the sea bed for geophysical research. They can also take photographs and collect rock samples for analysis.

▶ Manganese nodules can be dredged or sucked from the ocean floor. But most lie under 3 km (1¾ miles) or more of water. Engineers have not yet solved the problems of manoeuvring a ship and its collecting gear in open and often stormy seas. Furthermore, international lawyers are still debating the ownership of the ocean floor and its mineral wealth.

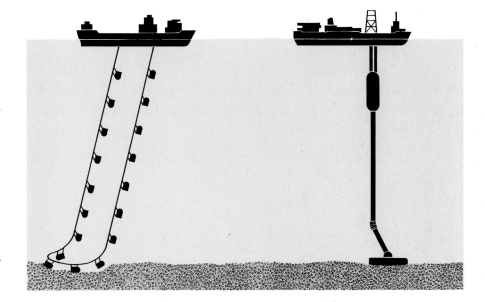

PREDICTING EARTHQUAKES

Earthquakes are a natural response to movements of the Earth's crust. Where layers of rock are either gradually stretched or compressed they can deform into basins or folds. If the strain is sudden, or too great, the rock fractures and shifts abruptly. In mountainous or volcanic regions the landslip caused by the fracture may be only a few metres long. Where continental plates are colliding or rubbing past each other the slippage may be more than a thousand kilometres long. The earthquake zones of California and the Middle East are lines of multiple slippage caused by plates rubbing past each other. The movements usually take place several kilometres below ground but the shocks spread through the earth and across its surface.

Few people are hurt by the earthquake itself; it is collapsing buildings that cause damage and death. People are buried in the rubble. Fractured electricity and gas mains bring risk of fire, and if the water mains and roads are damaged rescue services cannot get there to help. We cannot hope to stop earthquakes, but if we could forecast them, villages and cities could be evacuated, saving many lives.

Clues from the rocks

Engineers use sensitive instruments to measure the strain in rocks. But this is not a very practical method of forecasting because rocks vary so much in strength and structure that it is impossible to be sure when they will give way. It seems that many small slips along a fault line release stress and thus prevent a major earthquake. Lines of strain-gauges (see illustration opposite) have been set across fault lines in California, for example, to see if the strains are building up to danger level.

Scientists have discovered that areas about to suffer earthquakes sometimes provide other clues. Just before the deep rock gives way its crystal structure may open and close, releasing radon gas normally held there. Groundwater carries this gas into wells. Monitoring well-water might therefore give advance notice of rocks about to give way. These preliminary adjustments in the highly-stressed rock also change the electrical conductivity in the ground, which is easily measured. They may also be the source of electrically-charged gases which rise from the ground surface to give the faintly-glowing 'earthquake lights' that are reported from time to time. But none of these effects are universal, and they depend on the nature of the rocks under strain, and the volume of groundwater as

▼ Deep earthquakes are generated along colliding plate margins. Shallow tremors are caused by settlement and adjustment of the plates, by mountain building and volcanic activity.

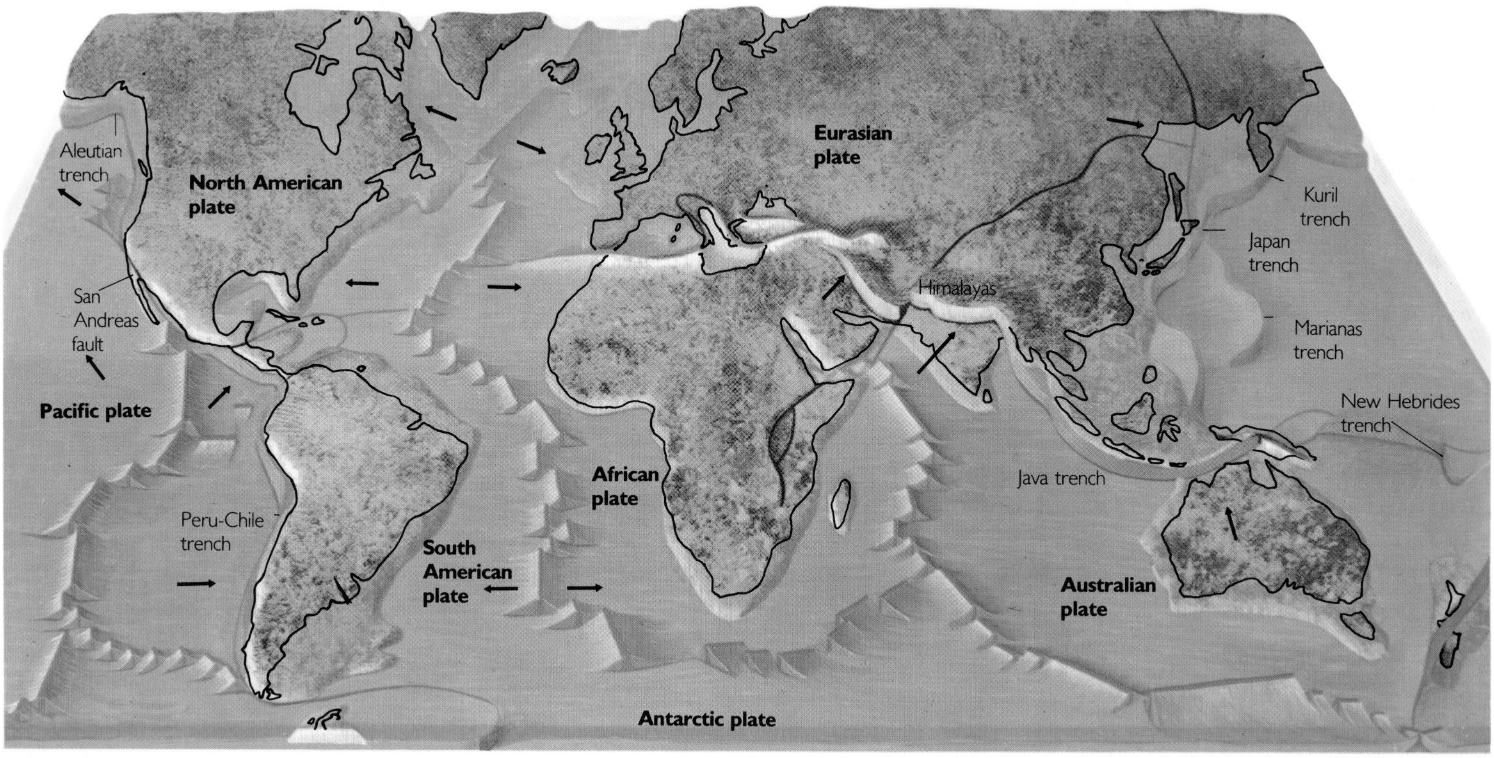

Aleutian trench

North American plate

San Andreas fault

Pacific plate

Peru-Chile trench

South American plate

African plate

Eurasian plate

Himalayas

Kuril trench

Japan trench

Marianas trench

New Hebrides trench

Java trench

Australian plate

Antarctic plate

► Most lives are lost when buildings collapse. Buildings can be made more resistant to shaking in earthquake-risk countries, but these techniques cannot save ancient buildings or those whose owners cling to traditional building materials and methods.

well. Clearly these are not reliable predictors.

A slight rising or sinking of the land surface sometimes takes place before an earthquake. Near active volcanoes this may be due to movements of *magma*, before it is expelled from the vent. Surveyors can now measure these movements extremely accurately, using laser beams from satellites to measure changes of just a few centimetres taking place over short periods. But much work has to be done before scientists can be certain that the movements will be followed by earthquakes.

Geologists in California are studying the fault system there in the hope of finding some reliable forecasting methods. Meanwhile, until our knowledge improves, we will have to rely on sensible building practice to reduce earthquake damage, because some districts close to active volcanoes and at risk from earthquakes have such good climates and such rich soil, that people will choose to live there, regardless of the dangers.

▼ These are some of the ways in which scientists try to predict earthquakes.

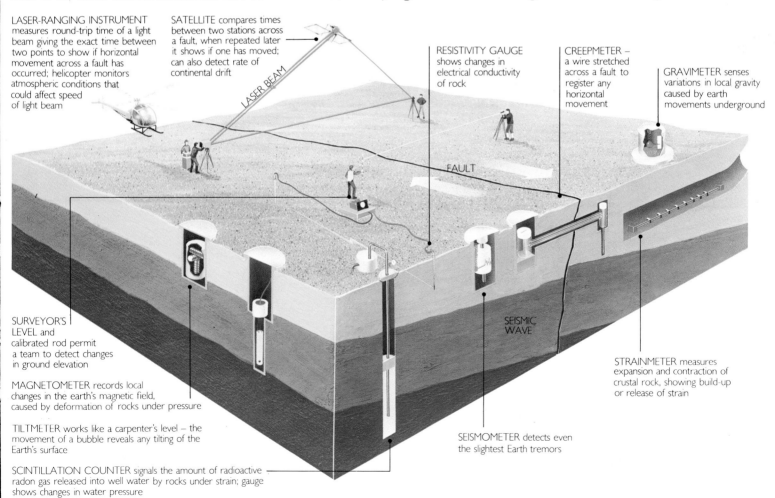

LASER-RANGING INSTRUMENT measures round-trip time of a light beam giving the exact time between two points to show if horizontal movement across a fault has occurred; helicopter monitors atmospheric conditions that could affect speed of light beam

SATELLITE compares times between two stations across a fault, when repeated later it shows if one has moved; can also detect rate of continental drift

LASER BEAM

RESISTIVITY GAUGE shows changes in electrical conductivity of rock

CREEPMETER – a wire stretched across a fault to register any horizontal movement

GRAVIMETER senses variations in local gravity caused by earth movements underground

FAULT

SEISMIC WAVE

SURVEYOR'S LEVEL and calibrated rod permit a team to detect changes in ground elevation

MAGNETOMETER records local changes in the earth's magnetic field, caused by deformation of rocks under pressure

TILTMETER works like a carpenter's level – the movement of a bubble reveals any tilting of the Earth's surface

SCINTILLATION COUNTER signals the amount of radioactive radon gas released into well water by rocks under strain; gauge shows changes in water pressure

SEISMOMETER detects even the slightest Earth tremors

STRAINMETER measures expansion and contraction of crustal rock, showing build-up or release of strain

LEARNING MORE ABOUT GEOLOGY

Geology is a science open to anyone with an enquiring mind because it can be studied at many levels. If you are interested in plants then geology will tell you more about the rocks and soils they grow on. If you are curious about the animal life in each continent, then geology will explain their distribution – why they are where they are. If you have read about past ice-ages or great tropical forests in many parts of the world, then geology will set these events in their correct order. There are plenty of books of course, but these are no substitute for the real thing. The best places to begin are 'in the field', at school, and in the museum.

In the field
There is no need to go far. Even if you live in a large city there are possibilities. As you travel round, look at rocks exposed in road and railway cuttings. Your home may be of local stone or slate but the major buildings and bridges may be of a different stone, tougher or more decorative. Out of doors look for fossils weathering out of limestone parapets and steps. Indoors, look carefully at polished hard limestone floors in colleges, cathedrals and town halls. They often show beautiful sections of fossil shells. There may be rocks in your local park, and don't forget the tombstones in the churchyards. As they are dated, you can get a good idea of the rates and processes of weathering of granite, marble or other ornamental stone.

In school
Collect small slivers of rock or fossils, from the garden or park if there is no open country nearby. Look at these under the microscope. You may be able to carry out simple analyses, to see if the rocks contain lime or iron. Collect and wash sand grains. With a hand-lens or microscope, see if these are fresh and sharp, or dulled and rounded. This will tell you if they have travelled a long way from their source rock or not.

In the museum
Familiarize yourself with the local rock types and fossils shown in your museum. How many ages of the Earth are represented in your area? Are there any mineral resources beneath you? How do these correspond to past or present industries in the district?

Further afield
If you can get out into country where rocks are visible you can hunt for fossils yourself. Take a good lens; many fossils are small and with the lens you will learn more about the rock types than you can see only with your eyes. Take a compass to see if the rocks are magnetic. If so, can you tell if they are volcanic or rich in iron? In hilly areas, can you trace the folds and slopes of the strata that form the mountains?

Past and present
We see the Earth at one moment in its long history. Everything about it – core, mantle, crust, oceans and atmosphere – was different in the past and will change again in the future. We can only guess at that future, but the past and present are laid out for our inspection. All we know about our planet has been discovered by careful study, here and now, but there are plenty of questions left for you to answer.

▶ Rocks as rich in fossils as this one are uncommon. From the beautifully preserved impressions of both the outside and inside of the shells of this creature scientists know the rock was formed during the Jurassic period of the Earth's history. This period began about 190–195 million years ago and ended about 136 million years ago. Scientists also know that during this period the climate was warm and humid and the shallow seas held many different creatures.

◄ Rocks and minerals can be extremely decorative – and not only when they are used in expensive jewellery! Throughout history polished marble, in particular, has been used as decoration in large buildings such as palaces, temples and churches. Notice how the architect of this church has used different coloured building stone to add interest to the interior of the church, and not just on the floor either.

GLOSSARY

Alloy A combination of two or more metals. Bronze is an alloy of copper and tin.

Chert A glassy mineral made of silicon and oxygen. It usually forms as slabs or blocks filling cracks in limestone.

Clast Geologists' term for the fragments of broken rock which go to form another sedimentary rock. These fragments may be as small as clay grains or as large as boulders.

Continental plate A block of the Earth's crust underlying a continent. Made mostly of granite, the plates are 33 km (20 miles) thick on average but thicker under mountain ranges.

Continental shelf Gently-sloping edge of a continent, covered by shallow water. Not all continents have shelves right round their edges. For example, South America has a shelf only on its eastern side.

Convection cell A portion of the Earth's mantle, heated from below, that behaves like liquid boiling in a pan. It rises very slowly, spreads, cools and sinks, to be reheated and rise again. There are several of these cells within the mantle.

Core An inner solid sphere, surrounded by a liquid sphere, at the centre of the Earth. The core is made of denser rock, perhaps with more iron and nickel, than the mantle.

Cracking The breaking of the complex organic molecules of oil into simple molecules. Deep underground, the Earth's heat and pressure causes the process to occur naturally. The same cracking process is carried out artificially in oil refineries.

Crust The outer shell of the Earth. It rests on top of the mantle. It varies from 5 km (3 miles) thick under the oceans to as much as 60 km (37 miles) thick under mountain ranges.

Crustal plate A block of the Earth's crust. It may underlie an ocean or a continent or part of each.

Element The simplest type of chemical substance with atoms all of the same kind. An element may be gas, such as oxygen, or metal, such as tin, or some other solid, such as sulphur.

Erosion A process where ice and rain eat into the rock surface and carry away the fragments.

Fissure A split or crack in rock or sediments. It may be so thin that water can just penetrate, or so large that lava or other minerals can fill it.

Geomagnetism The Earth's magnetism, which originates in the core. It is the force which turns a compass needle north-south.

Geophone A small listening device put into or on the ground, to pick up echoes from explosions on the surface. Used by prospectors, to discover how deep the buried layers of rock are.

Geophysics Study of the Earth and its atmosphere by physical methods. It includes study of earthquakes, magnetism, and gravity. Geophysicists are those who study these things.

Gravimeter A sensitive balance which can measure the force of gravity anywhere on the Earth.

Gravity The force responsible for the weight of any thing is equal to gravity. Gravity holds the water in the ocean basins and draws down the heaviest rocks towards the centre of the Earth.

Hydrocarbons Compounds of hydrogen and carbon. They come from buried remains of plants and animals and are mined as fuels. Coal, oil and gas are hydrocarbons.

Magma Hot molten rock, mixed with gases, beneath the Earth's surface. Magma may pour out at the surface as lava, or solidify below ground as basalt.

Magnetometer Instrument which measures the strength of the Earth's magnetic field.

Magnetosphere Region far above the atmosphere which is still influenced by the Earth's magnetic field, but is also blown into a teardrop shape by the solar wind.

Mantle The middle layer of the Earth,

▼ These large rocks were once trees. Now these fossils are all that remain of the forests that once covered the area.

▲ These beautiful crystals of fluorspar are the source material for the element fluorine, used in industry. Fine specimens of these crystals are often sold as ornaments.

between crust and core. On average it extends from 5 km–2900 km (3 miles–1,800 miles) below the surface. The rocks of the mantle are solid, apart from a molten layer at the top, but the heat and pressure are so great that they can flow slowly.

Mineral Any substance in the Earth which has a definite chemical composition is a mineral. For example, granite is a mineral, so is iron ore. Soil, which has no set composition is not a mineral. The word is often used to refer to the economically valuable products mined from the Earth.

Molecule The smallest portion of a substance which can exist naturally. It may be two or more atoms, of one or more elements. The oxygen in our atmosphere consists of molecules of two oxygen atoms. Water is a molecule of one atom of oxygen joined to two atoms of hydrogen.

Native metal A pure metal occurring naturally within the Earth. Gold, copper, silver and mercury may be found as native metals.

Ore Mineral containing a metal (or a non-metal such as sulphur), from which that metal can be taken, by some industrial process.

Pipe A shaft leading from a magma source, along which molten rock passes. The most famous are the diamond pipes of South Africa. Long ago, molten rock came up these pipes from deep within the mantle, bringing up the diamonds which

are now mined from it.

Plate margin The edge of a crustal plate. At constructive plate margins new plate is being formed. At destructive margins plate is destroyed as it is carried down into the mantle.

Plate tectonics The theory, now generally accepted, that the Earth's crust is made of thin rigid plates. Mountain ranges are formed where these plates collide.

Porous A rock or sediment with loosely packed grains, allowing water to filter through.

Primary ores Ores that have usually been formed by deposition from mineralized water deep within the crust.

Prospector An explorer hunting for useful minerals at or below ground level.

Salinity The relative amount of salts dissolved in water. In the open sea salinity is about 3%–4%.

Sediment Particles of rock or soil transported by ice, wind or water, and deposited away from their source.

Seismic belt Line across the globe along a plate margin, marked by earthquakes and active volcanoes.

Seismic wave The waves generated by earthquakes or explosions that can travel through or around the Earth.

Seismograph Instrument which detects seismic waves, and records their time of arrival and their strength.

Silt Sediment particles with a grain size of 20–60 microns. Silt is finer than sand, but coarser than clay.

Smelting Process of heating an ore to extract its metal content.

Solution Chemical weathering where minerals are dissolved by water.

Stable element One that is not naturally radioactive.

Stratum (plural: **strata**) The smallest division of layered rock; one distinct sheet or layer, laid down without a break. Strata vary in thickness from a few millimetres to many hundreds of metres.

Vein A fissure or crack filled with a different mineral from its surroundings.

Vent Underground passage from a magma chamber to the surface. Lava, ash and gas pass up it. When cold the vents of many volcanoes become plugged with soil and vegetation washed in from the surface.

ERA	PERIOD	EPOCH	DURATION	Principal events
CENOZOIC ERA 0 to 63–64 million years ago	QUATERNARY	HOLOCENE PLEISTOCENE	about 2 million years	The Ice Age; glaciers and ice sheets cover Canada and N Europe.
	NEOGENE	PLIOCENE	about 5 million years	Early man living in Africa. North Atlantic continues to widen.
		MIOCENE	19 million years	Africa moves against Europe, folding up of the Alps. Man-apes in Africa. Less forest, more grasslands; large mammals plentiful.
	PALAEOGENE	OLIGOCENE	12 million years	Widespread lava flows in USA, India, Asia and Europe.
		EOCENE	16 million years	North Atlantic opens to the Arctic Ocean. Various mammals evolve.
		PALAEOCENE	10 million years	Greenland, North America and Europe divide.
MESOZOIC ERA 63–64 to about 225 million years ago	CRETACEOUS	LATE CRETACEOUS	35 million years	South America breaks away from Africa. Flowering plants and modern trees appear. Many reptiles die out. A high sea-level world-wide, with shallow seas on many continents. Chalk laid down.
		EARLY CRETACEOUS	36 million years	
	JURASSIC		about 57 million years	Dinosaurs and other large reptiles dominant. First birds appear. North America moves away from Africa.
	TRIASSIC		about 33 million years	First dinosaurs. Large marine reptiles common.
PALAEOZOIC ERA about 225 to about 570 million years ago	PERMIAN		55 million years	World-wide lowering of sea-level. Inland seas evaporate, salts and potash laid down over much of Europe and North America. Deserts cover Britain.
	CARBONIFEROUS	SILESIAN	about 45 million years	Amphibians and sharks. First reptiles appear. Insects common. Many types of reef-building corals and shellfish. Volcanoes in Scotland. Large river delta lays down future coal deposits in England and north-western Europe.
		DINANTIAN	about 20 million years	
	DEVONIAN		about 65 million years	Plants and animals colonise the dry land. Large mosses, ferns and seed-bearing plants, insects and spiders. Britain lay south of the Equator in the desert belt.
	SILURIAN		about 35 million years	All life still in the sea, but a variety of coral, shellfish and fishes.
	ORDOVICIAN		about 90 million years	Volcanoes in many parts of Britain. Some corals, shellfish and soft marine creatures.
	CAMBRIAN		about 40 million years	Simple corals and invertebrate marine creatures.
PRE-CAMBRIAN ERAS about 570 to about 4,000 million years ago	The duration of the interval between the Earth's formation and the beginning of Cambrian times was about 4,000 million years. The Moon and the meteorites have an age of 4,600 million years and there is much indirect evidence that the Earth is also of this age. It is commonly thought that this date marks the differentiation of the Earth (and Moon?) into core, mantle and crust. The earlier period, in which the matter from which the earth is made was accumulating, may have taken place perhaps over a few tens or a very few hundreds of millions of years.			Soft-bodies creatures, algae and bacteria. Life began, probably more than 3800 million years ago. Some types of iron and diamond-bearing rocks seem to date from this early period and are not known to have formed later. Pre-Cambrian rocks are exposed in Canada, Britain, Scandinavia, India, Africa, Brazil and Australia.

Time chart showing the geological history of the Earth. The left hand column shows how scientists divide up this history and how they believe each era, period or epoch lasted. The column above gives the principal events that occurred.

Acknowledgments

Alcan, Architectural Association, Ardea, Australia News and Information Service, de Beers Collection, Carlo Bevilacqua, Carlo Bevilacqua/Sandro Prato/Mirella Bavestrelli, A Borgioli/G Cappelli, British Museum (Natural History), Canada House, Nino Cirani, Bruce Coleman, Douglas Dickins, Foto Dulevant, ENI, EPS, Gilardi, Giorgi Gualco, Robert Harding Associates, SE Hedin-bild, Alan Hutchison Library, Archivio IGDA, Institute of Geological Sciences, Keystone Press Agency, G Leonardi/G Pinna, Photo Loic-Jahan, P Martini, Museo Civico di Storia Naturale di Milano, NASA, National Coal Board, Peters, Picturepoint, Foto V Radnicky, Rapho, A Rizzi, SEF, Smithsonian Institution, Spectrum Colour Library, Titus, Vulcain-Explorer, AC Waltham, Woodmansterne/John E Grant, Diana Wyllie, ZEFA.